Author photograph by Peter Rutt

Stephen Rutt is a writer and naturalist. His is the author of *The Seafarers: A Journey Among Birds*, which won the Saltire First Book of the Year 2019, a Roger Deakin Award and was longlisted for the Highland Book Prize; *Wintering: A Season with Geese*, one of *The Times*' best nature books of the year for 2019; and *The Eternal Season: A Journey Through Our Changing British Summer*. His writing has been published in the *Guardian*, *Scotsman*, *Sunday Post*, *British Birds*, *EarthLines* and *Zoomorphic*. He lives in Dumfries and Galloway.

The Waterlands

Follow a raindrop from source to sea

Stephen Rutt

Elliott&Thompson

First published 2026 by
Elliott and Thompson Limited
2 John Street
London WC1N 2ES
www.eandtbooks.com

Represented by:
Authorised Rep Compliance Ltd
Ground Floor, 71 Lower Baggot Street
Dublin, D02 P593
Ireland
www.arccompliance.com

ISBN: 978-1-78396-931-9

9 8 7 6 5 4 3 2 1

A catalogue record for this book is available from the British Library.

Typesetting: Marie Doherty
Printed by Pixel Colour Imaging Ltd, 10 Prestons Road, London E14 9RL

For A, for everything

Contents

Somewhere before the beginning of the event – the fall of a raindrop – lies its real beginning. Plant roots tap the rainwater from the soil, pumping it around their flesh, storing it until spent. Sated, they exhale, releasing droplets of water vapour back into the air. The word for this is transpiration, its origin in a Latin compound of the words for 'through breath'.

The hot breath of the earth helps too. The sun loosens molecular bonds, charging water – in puddles, ponds, bogs, marshes, rivers, canals, oceans – with energy, making the surface molecules dance, coaxing them into the air, into evaporation.

Rising high, it builds into an invisible sea in the sky that grows heavy, slowly. Vapour droplets coalesce around grains of dust, condensing back to liquid water. The cold, eddying atmosphere turns the droplets to ice or jostles them, merging as they collide, growing heavier.

Clouds are laden now. The sea in the sky made visible. Water drops, created from a million small vapour droplets, are waiting for the right moment: when the clouds hide the sky in darkening shades of grey and the first few – the heaviest – begin to slip from the sky and fall. A threshold is reached. The balance tipped. Suddenly they all begin to go. The water returns to the earth as rain.

The raindrop's journey. Now we can begin, again.

Introduction: Before the Fall

When I was two years old, I fell into a river.

It was my mother's birthday. We were walking through Cambridge and I saw a flower growing out of the bank of the River Cam. In the family story, I wanted to pick it for her. Before my parents could react, I had pitched through the overhanging bank, into the river, cheated by land that wasn't solid. For a moment all was dumb – the absolute suddenness of surprise cold water – and I remember the light of the sky and the feel of the water and then being heaved out by my frantic father, sodden and shivering with shock.

It is one of my earliest memories.

It would be cheap to say that is why I can't swim. But I can't swim, and not for want of trying. I merely thrash about with unsynchronised limbs, chlorinated pool water stinging; feeling the shame of my body, the shame of being the worst in school; never able to take that step of faith into the deep end. It is as alien to me as dancing. In the shallow North Sea I can wade for a distance up to my knees before the anxiety rises. At midriff I have to stop. Cold water grasps with a tight grip. It makes me gasp and suck at air that feels like the asthma attacks that plagued my childhood. I fight breathlessness as I stumble back to a warm beach. My attitude is like the fisherman of folklore who thought it was bad luck to learn to swim; or like Gavin Maxwell, another non-swimmer, who thought the Mesopotamian marshes 'as good a place to drown as any other'.[1]

My daughter, a toddler, has been in a swimming pool more times in the first two years of her life than I have in the last twenty years of mine.

And yet: water holds me. Water holds my mind and my heart. I am drawn to be near it. To walk beside the river, to seek out the loch on the map and catch a glimpse of it through tree trunks, the ground a mirror of the sky. I learned to birdwatch by wetlands, holidayed by lakes. Popular anthropology suggests that we are apes who evolved to be near water. The first records we have of life on Earth – nearly 4 billion years ago – are bacteria that were laid down in shallow seas and are readable in rocks.

Water is life. The cells in your body – in all bodies – rely on water to work, to hold their structure and use oxygen. Everything on Earth requires water to live. All life beyond Earth probably relies on water to live. At least we think so: when we scope the far corners of our galaxy, our satellites spinning through space are searching for the signs of water. We don't know how it is that Earth stands alone in the vastness of space as the only wet planet – perhaps some miracles are destined to be unexplained – but we do know it is essential. Every glass of clear fresh water that we pour from our kitchen taps is an example of a miracle that sustains and explains life on Earth. We can't conceive of life existing without water as its animating magic. The science writer Alok Jha puts it this way: 'Every living thing is simply a different inflection of water; a deviation of a few per cent from purity.'[2]

I have forgiven land for the trick it played on me as a toddler. But I now walk carefully towards riverbanks, holding my daughter's hand tightly, aware that the ground might not be so solid. I stop at the boundaries of the marsh, where the surface can be quicksilver and self-willed; happy to wait by the lapping of the lochshore while others swim. While they let their bodies be held by water, I am happy to let it hold my mind.

I take a breath near the top of Craigie Fort, where the rock levels out, after a short, steep climb. Loch Lomond, Scotland's biggest body of water, glimmers just beyond, through the silver-brown lacework of trunks and branches, above a rusty tangle of bracken. I had been nurturing a quixotic wish. After a housebound winter, I was chafing at the constraints, bouncing off the walls; I wanted to see the loch as a whole, the entire 36-kilometre length of it, shaped like the point of a prehistoric spear, thrust into the Highlands. I wanted to see the space within its limits, distance defined by a rim of rock and not the unconfined vagueness of a coastal horizon.

A few steps bring me to the edge of the wood. Land drops away. Below my boots, the wet grass of winter bends down to the bare canopy of the trees below, but it would be perverse to focus on your feet here. Loch Lomond is spread out, a vast carpet of water bathed in winter sunshine on this still morning; its surface polished by the sky into a brilliant blue, backed by the emerald and russet-flanked mountains to the north – Ben Lomond, Ben Vorlich and Ben Vane, the highest peaks in this corner of the Highlands – and the gentler grass-green hills to the south. A scattered line of islands runs west of here, their names on the map reading like an incantation: Inchcailloch, Inchfad, Inchcruin, Inchmoan, Inchconnachan; these Inches are the lengths of land left by the ancient distortions of rock that made this place, that let water droplets collect and pool in this loch.

A cormorant on a post leans cat-like into the sun. Great tits sing like a hinge in need of oiling. Chaffinches flit through the firs in a frenzy. Swimmers yodel as they walk into the water; dry-robed shiverers stand on the shore. The view is good. Famously good.

You might even say 'bonnie' and sing a song about it. The space and light and shine of the water is reviving to my senses after a long grey winter in a small flat. I breathe it in.

In *Belonging* bell hooks wrote, 'The idea of place, where we belong, is a constant subject for many of us.'[3] Water, one of my constant subjects, draws me deeper. It is not a place, but one of the building blocks of the many types of wetland; they vary depending on the water – its motion, its chemistry, its interaction with a location. Water has a profound effect on the landscape, shaping its appearance, the nature it holds, how we interact with and respond to it. Its subtle variations give light and shade to a place. Rather than tying me to a location, it allows me to travel, taking my fascination with habitat with me. There is water (almost) everywhere. It joins land to sea and sky; life to rock; past to present; climate to weather.

A wetland simply means a habitat where the land is flooded by water regularly. Ramsar, the international convention on wetland conservation, classifies them as places of water no deeper than six metres at low tide. Wetlands are normally categorised not by the water but by the type of vegetation that grows with it. I'm no botanist, so I'd rather define it by the play of water and land, and the permeable border between the two.

Water and land are not separate, nor are they two sides of the same coin, but more profoundly intertwined. They are the blood and bone of the earth. So I call them waterlands, partly as a nod to the Graham Swift novel and partly because a looser definition better fits the unfixed nature of water. Any place where water and land join together is where an alchemical sort of magic occurs. Life blossoms out of the mix of rock and mud and water; the sort of ever-shifting, ever-changing kind of life. Heraclitus, in 500 BCE, is supposed to have said, 'No man ever steps in the same river twice.' Or as Mr Crick, the narrator of Swift's *Waterland* has it, when he describes the process of silt in the fenland rivers, it 'shapes and undermines . . . demolishes and builds . . . is simultaneously accretion and erosion; neither progress'.[4] In the waterlands, change is a given: things are always in a state of flux and flow, always being done and undone at the same time; as liquid, water performs a give and take with land, scrubbing away at riverbanks and building marshes. Water is never still, even in a lake; it is withdrawn and deposited elsewhere, the water cycle feeding and refreshing the waterlands.

It is a day for the dance of evaporation. By the shores of Loch Lomond, the light has lost its lustre but the sun retains its strength; the far hills fading out in a soft glaze, the day's gentle greying of the landscape. Water molecules are in constant motion with each other. Increasing their energy increases evaporation: days of brilliant light and warmth – or short-wave and long-wave radiation, if you wish to be technical – energises the molecules. Humidity holds the process back (the more water in the air, the less it can take up). The lack of wind does not help stir the molecules further but instead leaves what is there to hang, hazing the horizon. In a few months' time the evaporation will be increased by the stomata – the pores of leaves – opening, exhaling their vapour from Loch Lomond's famous oak woodlands.

This is the story of the water cycle. Our water is a gas now, leaving land by evaporation or transpired by plants. It moves, swept by the atmosphere's currents, cooling as it rises. NASA knows what percentage of Earth's water is present as vapour in the atmosphere at any time. It is 0.001 per cent. A thousandth of one per cent. This is vast compared with the percentage of Earth's water that is in rivers: 0.0002 per cent. Most of our planet's water, 96.5 per cent, is in the seas and oceans of the world.[5] The evaporation happening here from Loch Lomond is just a slither of the whole; this cycle is driven by the oceans (86 per cent of Earth's evaporation happens at sea).* Their vapour drifts overland and after an average of nine days cooling and condensing around tiny particles high in the atmosphere, it reforms into droplets, before falling as rain and refreshing the earth. The loch's evaporation joins the great gassy sea in the sky, moving the lochwater elsewhere in the world. The rain that falls on Lomond is more than likely Atlantic water that has shed its salt on evaporation.

Water is a closed system. It will forever be water – never destroyed, only transformed. Each droplet is forever trapped in the water cycle, repeating each step for eternity, like Sisyphus: becoming vapour, circling the world on atmospheric currents, condensing and falling back to the earth's surface, flowing as a liquid or frozen for a time in an alpine glacier or Arctic ice cap, changeable but ultimately indestructible.

Rain is what sets the whole cycle of the Earth working, the most visible way that water begins to move through the landscape, charging it with energy and potential and feeling. It's the same here in Scotland – a self-professed 'hydro nation', rich in water and with plans for it – as it was in the arid east of England where I'm from, when black summer storm clouds would arc over dusty fields; the same as it was in Hungary and Greece, when thunderstorms caught me out on holiday, turning streets into shallow rivers in a single flash of lightning.

When it rains here, the clouds release water straight to the surface of the loch, but it doesn't need to rain anywhere near the loch for the water to end up here. The Endrick Water is a river that begins 50 kilometres away as a series of hill burns braiding together in the Gargunnock Hills, southwest of Stirling, before flowing into the

---

* The discrepancy between these figures is because most evaporation happens in the warm subtropics and much of the ocean is either too cold for efficient evaporation or being rained on.

southeast corner of the loch. The slopes of Ben Lui to the northwest, the high hairpin of the Coire Earb to the northeast, the bare hills above Faslane, well to the west: rain that falls in all of these places will eventually flow through Loch Lomond. If it doesn't land directly in one of the rivers, it can end up in one of burns that become a transient torrent, a white scar falling straight down a cleft in the steep hillsides; or in the bogs that form in the hills where *Sphagnum* mosses soak up the rain, like a sponge that slowly releases its water over time, keeping the hills wet; or in the RSPB-managed fen here, where rain meets surface and ground water, and the unique plants and insects of the waterlands flourish. It could end up trapped in dammed – damned – Loch Sloy, behind the walls of the hydroelectric plant or be allowed to trickle through the Inveruglas Water to join Lomond when we decide.

We have been using and manipulating the water cycle for our ends for as long as we have been a species that settles and dwells, from the 12,000-year-old stone-cut cisterns of Göbleki Tepe in Turkey, which collected rain water in one of the earliest villages we have found, to the basalt-black deserts of northern Jordan, where our oldest surviving evidence of damming comes from the fourth century BCE; in the Indus Valley of Pakistan, drains, sewers and toilets existed by the second century BCE; roughly the same time as the Mesopotamians had constructed a series of irrigation ditches for channelling water around the fields that fed one of the great early civilisations. Because water is never lost, only transformed, it is plausible that one droplet in the cycle could have visited each of these civilisations in turn. Collected in Turkey, drunk in Jordan, flushed in Pakistan, nurturing the barley near Basra; coming out of your tap for your next drink.

Only transformed. Water, nurturer of civilisations, is becoming an increasingly contestable thing: threatening in both its surfeit and deficit – and quality. The former vice president of the World Bank said that if the twentieth century's wars were about oil, the twenty-first's will be about water.[6] In 2022, the east of England was unrecognisable: parched and officially experiencing a drought, while the west of Scotland remained resolutely wet. It is a pattern repeatedly globally. The American west is experiencing a 'mega-drought' – its driest conditions in a millennium[7] – while in 2023 Fort Lauderdale in Florida was hit by 63 centimetres of rain in twenty-four hours.[8] The year 2022 was Australia's ninth wettest on record, with flooding a regular, devastating occurrence, while Australia also equalled its hottest day on record of 50.7°C.[9] The world is getting both wetter and drier as it heats up.

Only transforming. A crisis in water affects humans as well as nature. It shapes our present as it has shaped our past. Currently, in the war zones of Yemen and Ukraine, a resurgence of cholera is feared from desperate people drinking dirty water. In Flint, Michigan, nobody went to jail after lead poisoned the city water supply for years: toxic chemicals including PFAS and metals such as arsenic are also present in American tap water. In Australia, 'universal and equitable access to safe and affordable drinking water for all' has not yet been achieved, with 600,000 people drinking substandard water.[10] The head of the British government's Environment Agency sees droughts due to climate change as an existential issue.[11] Even here in soggy Scotland, where water and rain underpin much of the country's cultural identity, drought risk is rising. We are being asked to conserve a thing once bountiful. And it is vital that we do.

'Water is our common future,' wrote Audrey Azoulay, the director-general of UNESCO.[12] And yet it is a future we are frittering away. The Colorado River should flow into the Gulf of California in northwestern Mexico, yet the United States Geological Survey estimates only 10 per cent of its water ever makes it to Mexico.[13] How the USA and Mexico use the water of the Colorado River means it rarely ever meets the sea any more, severing the vital flow of life that comes with functioning rivers. Water is a shared resource:

evaporation from one area can fall as rain far away. Rivers bring water from one country to another. International cooperation is essential – but not always a certain prospect.

What we seem to have lost in the well-plumbed world is a certain reverence or respect for water. Modernity has brought it to our homes instead of us needing to go out and seek it. It flows out of our taps and showers: a miracle of modern life, health and hygiene, a liberator of domestic time and work. But it is harder for us to connect it to the water cycle, the great movement of water through the landscape and how it changes with the weather and climate. Now that we are not collecting it, pumping it, living life via a bucket, water has become almost abstract. The Scottish environmental historian T. C. Smout makes the point that in contrast with the culture of more arid lands, water here is 'not a treasure until it had been rarified. Uisge beatha – whisky, the water of life – came not from a fountain or a holy stoup but a bottle.'[14] Without that rarity we have frittered and wasted our abundance.

The relationship we have with water currently is strained. Like the slow-motion break-up of an old relationship, the familiarity and ease of it have bred not contempt but the potent rot of being taken for granted. Because without having seen the danger, or believed in the danger, one day we'll wake up and the water won't be there, or it'll be in the wrong place or polluted beyond use. If water begins to fail, life goes with it and we'll be left grieving what was and has gone, what we've let slip through our fingers.

Figures from the Wildfowl and Wetlands Trust (WWT) do not tell a happy story. We have been bloodletting: 87 per cent of the world's wetlands has been lost in the last 300 years – marshes drained, bogs dug, saltmarsh reclaimed, water meadows built on.[15] The letting has not abated. In February 2023 some of the canals of Venice ran impassably low in places. The Aral Sea between Kazakhstan and Uzbekistan, which was the world's fourth-largest lake in the 1960s, has been destroyed by irrigation along the rivers that flowed into it. While cottonfields grew, the lake dwindled to a quarter of its historic size. Water is life; its absence takes away more. As the water levels dropped, the lake grew brackish and briny. The fish died, the ground became poisoned, fertilisers and pesticides washed downstream concentrating, not diluting. Fishing boats slowly corrode in a desert that was once deep water.

It continues. A third of the world's wetlands have disappeared since 1970 and they are disappearing three times faster than forests. Much effort and energy are expended to replant woodlands – shorthand for environmental good (although it is frequently not) – but only specialists seem to want to make our land wetter, despite, according to the WWT's numbers, 40 per cent of Earth's species relying on wetlands.

It is possible to be profligate with abundance. We fritter water away, curse the rain that refreshes. Water companies are concerned with their economies, not our ecologies, and spill sewage with what feels like impunity; the fines they pay seem to tacitly say that throttling rivers, killing fish, poisoning swimmers is good business sense. In the rare spare moments of fatherhood, washing up or doing the laundry while everyone else sleeps, I have found a deep rage simmering inside me. I feel it for the injustices of the world and our damaged and depleted future, for all that is happening to the planet. To rage against my daughter's future inheritance of this world: a dried-up Venice, British rivers as open-air sewers, the lead-spoiled drinking water of Flint, Michigan. It is harder to be passively hopeful or gently mournful about the world now. Harder to go with the flow.

It is August in Galloway, so it is raining of course. It falls twice here. First into the trees, where it licks the birch leaves with a slick of glistening light. Then it is flayed from the trees by the wind or pulled down, the drips dropped by gravity. The rain is crackling like static. This loch is small and sheltered, the water a sort of peat-black mirror of pines and gloomy skies, normally only scratched by the reflections of flying ravens. But now it is bursting with crowns that appear for fleeting half-seconds, disappearing into widening circles that refract into other circles, a pattern breaking like an animated paisley that passes in seconds: refracts, reforms, disappears – a hypnotic cycle of patterns that keeps running for the duration of my time here.

I come to this loch to be alone, because it seems that nobody else comes here. I find myself nestled in the cover of the boggy edge – but not alone. A dragonfly, a four-spotted chaser clinging to a bog asphodel gone to seed, its crystal wings held by droplets, the soft down of its body tousled like wet fur, a reminder of our shared weather and our shared reaction, the need for shelter. Next to us a reed leaf hangs down, the resting droplets magnifying the fine corduroy texture of its fibrous skin.

It is a good place to let the mind meander, to look closely at things. I think about how if I came here to be alone for a year, collecting all the rain that fell on me, I'd have drowned in two metres of the stuff. Instead, it puddles then percolates through the peat. Evaporation will sweep up the excess from the loch over the next dry spell, or it will find the creases of the land and flow gently towards the nearest river, tracing the deep connections of the surface of our land, the increasingly extreme nature of everything.

Joan Didion, in her parched corner of California, came to 'think about water with a reverence others might find excessive'.[16] Not to Nan Shepherd, in the soggy northeast Highlands. She does not 'understand it . . . cannot fathom its power'.[17] Tristan Gooley, the navigator, mimics Heraclitus when he writes, 'We can look at the same stretch of water every day for a year and not see the same thing twice.'[18] Some days that seems appealing.

I understand the reverence. Because water is a loose thread. You can pick at it but it slips through your fingers. You can pick at it and unravel an endless set of thoughts and feelings, fancies and furies, power and despair. Between a raindrop and Loch Lomond I feel all of this too: the gulf between water on the scales of molecule and loch; this thing that stays the same and simultaneously flows into vast differences. It sweeps me up.

As with so many of the everyday miracles that surround us, we have become over-familiar with water. But now I want to rediscover it. I want to explore the wetlands, to know the way water moves through the landscape, bringing life, entangling other topics in its currents, to understand its hold over me better. I want to understand the water cycle, how it shapes the land, shapes our lives – and how we shape it in return. I want to follow water as it moves – to track a raindrop as it falls to the ground and travels through the landscape, from river source in the upland moors to river end in the saltmarsh-flanked firths and estuaries of Britain.

But the movement of water through a landscape is not a two-dimensional linear journey. Being a water droplet is like being in a game of snakes and ladders, moving in every direction. You get delayed, you leap forward. A bog soaks you up; you land in a puddle and disappear back into the sky; a fast river transports you straight to the sea; you spend a year in a loch. The reality of water's movements from sea to sky to land to sea is an intensely complicated web of directions and the luck of where the droplet lands. Each droplet is perhaps best considered as undergoing its own cycle, its own version of the same basic blueprint. Some renditions of it can be measured in hours, some in millennia (scientists in Canada, studying inside a deep mine, have discovered 2-billion-year-old water).

And so, as I follow my raindrop's progress, I too will digress into the watery habitats it feeds: into lochs that are suffering from acidification; blanket bogs that are simultaneously land and water, a thin skin of peat over millennia-old water, locked and bulging below the surface, trembling with each step; England's chalk streams, globally rare and with a unique assemblage of wildlife, relying on historic rain locked in aquifers. I will digress into close-up views of the vivid, startling life that inhabits these watery worlds, who can tell us so much more about them. And I'll consider the ways water interacts with us, and how we have affected its form and flow. Some might say water is no longer a purely natural phenomenon, that all water on Earth now bears our fingerprints, through industry, pollution, destruction and restoration.

From geography, ecology, climate change, natural history and social history, I want to absorb everything a single raindrop can show us on its journey from source to sea. To follow the game of snakes and ladders it plays through the environment, through rivers, lochs, bogs and marshes, to flow with the life it makes or facilitates. To understand the stuff better.

It all starts at the source.

For a second there is nothing.

Then you can see the surface tension of the water as it breaks, the slick halo around the head vanishing on the apex of the upstroke. A breath. Under, again, with a sinuous tail flick between the ripples. Two dives later it brings up a fish, something wee, clamped between its teeth like a second silvery tongue. A flick of its head and the fish disappears.

Then under. The otter is always facing forwards into the flow, its tail like a hanging pause, sometimes breaking the water of its own accord, as if the two ends have decoupled, its unseen limbs furiously paddling through yesterday's rain. I stare at the ripples – the play of light and shade across the moving surface – trying to decipher the stray burst of air bubbles, the slip of a back, the subtle breath.

After a few minutes it tires of fighting the current here, just below the weir, where the water runs with the noise of an untuned radio. It swims to the shingle island, where the straw skeletons of dead plants are still lying half flattened by the flood, and where a few black-headed gulls stand. Normally otters and birds don't mix well but today they seem content, unflustered by each other, as if in mutual recognition that land might be necessary but not really their element. The otter moves awkwardly on stones, its body low-slung, its limbs returned to their original evolution, the land mammal's original step: walking out of water.

There is always this contrast between elements. Call it a kind of surface tension. The otter slips back into the water and repeats: a dive down and out of sight, then a re-emergence to breathe that becomes a dive again in one perfect, fluid motion. The afternoon is dwindling away. The light turning golden, the sky deep blue, January's chill nibbling at my ungloved hands. On the far bank a few people have gathered to watch the same thing as me, lured by this presence; this animal that crosses the boundary, slipping from water to land to water to—

1

Unreliable Sources

It falls in a moment. When the heaviest droplets of ice can no longer be held, the first slips from the sky and plunges, down through the damp, cold air, thawing as it falls. Once one slips, the others come. The earth inhaling rain, again. Our raindrop splashes into the sodden soil of the hillside, reunited with its kind, no longer an individual but suddenly sticking together, part of something greater. It takes a second or two, in the chaos of ripples, before the raindrop feels the real forces; pushed from below, by the other droplets emerging from under the ground; then a pull, the tug of gravity and the droplet flows again, the rainfall merging with the river source.

'Just six miles north of here,' I say cheerfully as Miranda drives, headlights on, along a Moffat high street washed clear of people by the rain. There's still an hour until sunset.

A few minutes later our ears begin to pop with nervous gulps. The grey road snakes up through a grey sky that grows thick and begins to hang heavily. The valley looks vacuum-packed. The bare rocks have a skin of ice, a grey glistening under the low cloud that is gathering at the corners of the day. The winter-bleached grasses lie flattened. It's a sign of the weather that we are driving into, a sign that we don't notice. Then we stop being able to see the signs. Stop being able to see much at all.

The fields on each side of the road are patchy white, then, as the car climbs higher, what we can see thins. The snow gets deeper. There is nowhere to turn around. Another surprise sharp bend, another climb and the view left and right shrinks to the two pillowy stripes of snowy verge. Following the road ahead is like looking through steamed-up glasses. The beauty spot to our right passes unseen. Our speed slows further. Lay-bys lit by headlights look like flattened icebergs. We pull into one, tyres scrubbing ice as they come to a stop. I swear a small, secular prayer we'll be able to get them moving again.

Tweed's Well, the source of the River Tweed, is marked by a coffin-like lump of weathered red sandstone by the road. It is carved with place names and a relief of ripples and fishermen, bridges and flowers. A flurry of white and grey lichens grows across its side. Snow has settled on top, covering a representation of the river. Irregularly shaped white plaques look like patches of that snow driven into its side; snow that tells the history of the River Tweed, alongside poetry by the former poet laureate of Edinburgh, Valerie Gillies, promising me that the source stays green when all is white. And the sting in the tail: 'Please keep to the roadside at this wild birthplace. There is no public path into the field.'

The source. The water cycle describes a definite pattern but it picks it out in vague lines; it is sometimes difficult to trace its actions and motions. It begins with rain that flows into rivers – but the rivers still have to begin somewhere. A popular old rhyme – the sort that was handed down, parent to child in these parts – begins 'Annan, Tweed and Clyde, / Rise a' out o' ae hill-side'.[1] And while that applies poetic licence to basic geography, it contains a truth: that Scotland-defining water flows from this small, wild corner of the southwest. It comes in rivulets and wells, springs and boggy slacks in the hillsides. Sometimes the river comes from the ground as one. Sometimes it is a blending of two sources, sometimes the braiding of many burns. Each river begins and ends differently, each has something to say about water, its beginnings, and what it has done to the glacially made lumps of rock that are the Southern Uplands, where these three rivers rise.

And so a river source is a more definite thing, a starting point for the flow of water over (and under) land. It offers clarity for the motion of water, one of its starting points in a certain journey towards the sea. Which is why I'm here today, the start of my journey to visit all three water sources, starting with the Tweed. It is an obvious place to begin. If you can find it.

I slide back over the ice, cross the road and crunch into the snow-ploughed slush refreezing by the side of the road. Behind a wire fence a white field drops away into a thick fog. Water, water everywhere: in the field, the sky, the air, yet not the drop I came to see. I am standing on the verge of an A road not far from a motorway and yet stillness and silence reign. Because nothing living is stupid enough to be here today except us, and only I got out of the car. The source of the Tweed – the well – is out there under snow and behind fog.

I had mentally steeled myself to ignore the sign's warning. To hop the fence and brave the bite of water as it locates the holes in my old leaking boots with each step through the bog. But this is beyond bravery. I look back. Miranda is taking photos of me through the car window to send to people, convinced of my daftness. I am reluctant to give it up, but maybe this isn't the moment. Maybe this is failure. One last look at the few metres ahead I can see. There are sheep hoofprints in the snow and a stubble of dark grasses breaking through the crust of January weather. I head back.

The car wheels spin and crunch as it turns around. I grip tightly as sharp turns materialise out of thicker fog. Somewhere out there it is hiding a thin green thread through the snow, the waters breaking, the Tweed's slow birthing.

It's difficult to imagine, under its veil of snow and ice, how the water continues to trickle out. But I have learned recently that the way water molecules fit together means that water is at its densest at 4°C. Below that temperature and its structure becomes crystalline as it solidifies, trapping air: it is why snowflakes fall from the sky not snowdrops; raindrops not rainflakes. It is why ice is lighter than water and floats, the solid on top of the liquid. And so the source will be finding a way to flow slowly regardless, warmed by the earth and slipping between grass and ice.

This was all water once.

Spin time backwards. Split Scotland from England and fill in the gap with an ocean, the Iapetus. Populate the sea with hagfish, lampreys* and scuttling trilobites. Cloak the land with the first fungi and *Cooksonia*, a minuscule moss-like growth that was the first plant on land to transport water around its veins. This is the starting point: 400 million years in the past. Play it forward from here. The land that becomes Scotland crunches into the land that becomes England. In this tectonic mess, the prehistoric landmasses that collide destroy the Iapetus Ocean. The ocean crust is pushed and shoved under the landmasses but, as this happens, the rocks from the seafloor get scraped off. They are peeled like the skin of a carrot and piled high, left to compost into a newly solidified set of mountains: the Southern Uplands of Scotland.

The old seafloor is mostly greywackes and shales, sedimentary rocks made of old mud and silt. These are not particularly interesting rocks. However, they are fragile rocks, prone to splitting in

---

* The coelacanths that still swim today are just about to evolve. The hagfish and lampreys in the Silurian seas are similar but not exactly the same as the examples extant today.

layers. There is an aquifer below Moffat in the heart of the Southern Uplands and large amounts of rainfall above. The water held under pressure by the rock finds its way out, pushed through these fractures. These mountains gush water.

Though we are still aeons and catastrophes away from this all becoming water again, in the compressed time of geology – that 400-million-year blink of an eye – it also does not seem implausible. In the meantime, these hills – tall, rounded and steep-sided, dozing after their traumatic beginning – are the genesis for three of Scotland's major rivers, each beginning as tiny silver threads that push out rock and earth before weaving into something substantial.

Professor Peter Smith describes this area of the Southern Uplands as being 'like a bagatelle board'.[2] But instead of balls and pins, this bagatelle is played with rock and rain. Upland areas produce orographic rainfall: rain that is written by the hills. Droplet-laden clouds are pushed up by the nib of the earth and, in reply, the sky's wet ink falls hard and cold on the peat and grass. That old rhyme, 'Annan, Tweed and Clyde, / Rise a' out o' ae hill-side', is self-evidently wrong. Roughly 8 kilometres separates Tweed's Well from the Clyde's sources, splitting them across two different ranges of hills; the Moffat and the Lowther Hills, two different council areas. The Annan is in a third, nearly 2 kilometres from Tweed's Well and over 9 kilometres from the Clyde's source. Yet there is a truth to it. One rain shower here can land on the border between three watersheds, so water can end up in the Tweed's long slow eastward arc to the North Sea or in the Annan, flowing south to the Solway Firth, or in the Clyde, snaking north through one of the Southern Uplands' major clefts, before it kinks west at Glasgow, funnelling out into the Atlantic. The original author of that rhyme wasn't to know (the water cycle wasn't properly described until 1931) that instead the Annan, Tweed and Clyde all rise out of one large cloud.

These three rivers offer Scotland in a microcosm. The Tweed wends its way through the Southern Upland's valleys, past castles, bastles,* ruined abbeys and vanished palaces. It gave its name to the cloth that became symbolic of upper-class countryside conservatism whose wearers fish the river for its venerated salmon: picture the Tweed and you picture Scotland in stereotype, even though the last stretch of the river flows through England. The Annan slips south past a Victorian spa town, then through prime farmland to its end, at its eponymous former port town. It is a quiet, sleepy river. But even sleeping rivers awake, like the Annan did in October 2021, when two bridges were washed away after heavy rainfall led to the highest river level in fifty years.

The Clyde is something different. Wholly Scottish in spirit, 'Clyde-built' is a badge of honour. Clydesdales are horses bred for power and pulling. The river, likewise, has pushed and pulled, powering both the Industrial Revolution in Scotland and then shipbuilding, while passing through some of the most urban and some of the wildest terrain; veering wildly between the pure and the polluted, the old and the new.

Annan, Tweed and Clyde: all different but trace them back and they all begin the same: as cracks in a welter of 400-million-year-old seabed, starting in roughly the same place. Out of this wild, wet area, the water begins to flow.

Nan Shepherd wrote, 'One cannot know the rivers till one has seen them at their sources; but this journey is not to be undertaken lightly. One walks among elementals and elementals are not governable.' She was thinking of rivers such as the Aberdeenshire Dee, which begins life as a spring on Braeriach, 1,220 metres high on

* Like a fortified farmhouse, designed to thwart cross-border raiders.

the Cairngorm plateau. She is not thinking of the River Tweed, 385 metres above sea level and normally visible from the car, a short distance from a major motorway. And yet these places, self-evidently special, do seem to have their own minds, their own wills. It's not possible but it seems possible in the moment, that Tweed's Well knew my plans, and its elementals decided, 'Not today.'

Shepherd's next sentence: 'There are awakened also in oneself by the contact elementals that are as unpredictable as wind or snow.'[3] I am as spiritual as a doormat. I cede those experiences to other people. And yet.

Later, at home, I read that the Tweed has a track record on this. Walter Scott recalls with a wink the story of a local baron who, called away on crusade, returns after 'seven or eight years' to find his wife with a young son. She explains, through the words of Scott, that she walked past the Tweed, not far from here, when 'the tutelar genius of the stream' emerged from a pool and fathered the son.[4] A little further downriver one of the figures reputed to be the original Merlin of Arthurian legend is supposed to have haunted the woods of the upper reaches, a half-mad seer. Two metres of rain falls here every year, which is enough to unhinge anyone.

Maybe my trip to Tweed's Well wasn't failure. The airless archives of the internet – where the weather is always set fair – show me the well in several different moods. The National Gallery of Scotland holds two drawings by Sir George Reid from 1883. One in pencil, one in pen; both simple pools of paper surrounded by the shadows of shaded pencil or pen strokes, and the edges of delicately picked rush. Much has been left undrawn and the effect is of an open, peaceful, light-filled place.

Fifty-five years later Robert Moyes Adams took four photographs that are now in the St Andrew's University archive. A shepherd – the caption identifies him as a Mr Neil Manning, with dog – stands in the field. The wind tugs at his jacket, the collie looks off camera, as if with half an eye on the incoming weather. They are next to a small cairn of rocks that rings a pool, much smaller in appearance than in Reid's drawings. Each photo shows the same subject, sometimes with a second dog, but at a slightly different angle. They are as light filled as Reid's drawing but each photograph increases in bleakness. Adams was a landscape documentarian. These are not aestheticised images and because of that they possess a strange atmosphere that I can't entirely explain, as if the landscape was just miles of monochrome and molehills, dead grass and uncanniness. It seems impossible, out of the context of the place, to imagine this tiny puddle as being the beginning of one of Scotland's most fabled rivers.

I intend to return to the area but then I have to pause for Covid. The virus turns my lungs to dust, cancelling the walk I was going to do with a friend. The day before our rescheduled walk and the virus strikes him. It feels as though Shepherd's ungovernable elementals are still holding sway, still deciding for us: not today. I bow to the curse of Tweed's Well. I turn my focus to the next source, a kilometre or two to the south, where the River Annan gushes like a leak from the Devil's Beef Tub, on a 60-kilometre drift down towards the Solway, through unsung Scotland.

The Devil's Beef Tub is hidden. Lying below the crest of a ridge line, the bleached brown grasses merge with the hills behind, erasing the gap, making it feel like miles and miles of gently undulating moor, a vast wet space.

The Beef Tub is a corrie, a circular hollow in the hills carved out by a glacier – I prefer cwm,* the lovely Welsh word for the same phenomenon (you may also know it as a 'cirque'). Here four hills huddle around the top of Annandale, all steep flanked. It is alarming.

* Pronounced 'coomb'.

There is land beneath your feet. Then, a few metres on, there isn't. If your eye was focused on the hills and farms beyond, you might not notice until you end up like MacLaren, a Jacobite* who escaped capture by rolling down the hillside here.

There are two contested etymologies for the Devil's Beef Tub's name: the local Johnstone family were known as devils for their reiving, the cross-border skirmishes and cattle-rustling that plagued the medieval period. This is supposed to be where they kept their stash of illicit cows. The other suggestion is that it is a corruption of 'bath tub', with the sulphurous springs of Moffat giving it the requisite satanic whiff.

When I arrive, it's early afternoon, yet the light feels as if it is closing around me. The map marks the River Annan as beginning around here, not far from the road on the flank of Annanhead Hill. It would make logical sense. Water obviously can't fight gravity so it finds its way to the lowest ground in an area and I can see where it would be: a dip in the carpet of grasses, filled with a stubble of water-loving *Juncus* rushes, stiffer and greener than the grass. The ground towards it gets softer. Hair moss, emerald and starry, absorbs the shock of my boots. Water pools by my toecaps, trembling. I walk as close as I dare, which isn't that close. I might not be able to see the Devil's Beef Tub but I know it's there. I know the map's thick cluster of orange contour lines falls a short distance beyond this not so solid ground. The source here is vague, seemingly not formed from a spring but just a dampness, water weeping from the hillside. I turn around and notice that a drainage ditch runs alongside the forest behind, feeding this wet area, blurring the river's beginnings on a rainy day.

---

* Jacobites were supporters of the House of Stuart, the royal family deposed for being Catholic. Between 1689 and 1745 they fought a series of unsuccessful rebellions.

The Devil's Beef Tub wears the River Annan like a stethoscope. In the Annan's uppermost reaches one flow becomes two, each fork burying into the ear-like cleuchs* on each side of Annanhead Hill. To get to the east fork, hopefully a more definite source, you have to go over the crest of the hill, funnelled in without choice by a 250-metre drop on one side and a fence and a dyke on the other.

Drizzle begins, a damp kiss on each cold cheek. The way is straight and steep. A bare-peat and dead-grass trudge; step, step, slip; staring at the water pooling in the soft peat imprint of other feet. The air is not quite enough to refresh my lungs, the after-effects of Covid burning in my alveoli.

The top emerges and for a second I think I am hallucinating the bench and its promise of rest, like a desert oasis. A plaque reads: 'Bernie's View'. I am grateful to Bernie for a moment to sit and share in the landscape. To the north a black block of pines hides the view towards the Tweed. The eastward view is just hills, bare and softly rolling to the horizon, disappearing in drizzle (earlier I passed a group of planters carrying spades and saplings who were here to correct that bare emptiness, planting native trees for the Borders Forest Trust). Bernie's bench is a wobbly wooden thing, surreal for looking as if it belongs in a suburban park, when it is actually perched at 478 metres above sea level on top of a hill. It is tethered to the nearest flat ground to the edge, but to see the Beef Tub you have to stand up and creep forward as far as your personal tolerance for vertigo will allow you. Mine does not let me see very far.

The drizzle strengthens. I carry on, hoping the bulk of the hill at my back will offer some shelter.

It does not.

The next hill goes almost unnoticed. Peat Knowe, a small flat area on the descent is acknowledged by the map but not my feet.

---

* A Scots term for a rocky crevice.

Carrying on down the far side of it, a hollow knocking sound catches my ear. Water, pouring, louder with each step, until I'm ankle deep in the squidge of soggy soil at the top of the cleuch. A spring, as though the moor behind is a tap turned on, the weight of water in the peat pushing out by a rock and gushing like a series of miniature waterfalls for a few feet, soothing the moor grass to face downwards, away from the light; slick to the hillside like the devil's wet shower curtain.

In the notch between hills, *Cladonia* lichens grow in the grass like off-green golf tees. Orange lichens are splashed like false suns on the rocks. A red-leaved blaeberry sapling is finding a way to reach skywards out of all this soggy ground. I pause. There's a conundrum here: the gushing spring cascades down a contour, pooling in a visible wet patch – and then the water vanishes. This cleuch doesn't cascade with white water tossed from rock to rock. The map marks this spring as a single blue dot above a scree slope, and below that the spring that begins the east fork of the Annan, apparently unconnected to what we can see above ground.

The river as shown by the map is not quite all of it. Water finds the path of least resistance, the same one that all other water in the same position would take. Behind me the land plateaus between the hills and anything that falls there will end up in the watershed of the Tweed, even if the Tweed itself is invisible. It's a seam, a dividing point in the landscape – but it is a subtle one. Not the psychogeographer's liminal division or the pagan's sacred threshold; to hop the fence and dyke and squelch over the rough wet moor on this drizzly afternoon would not be revelatory, just uncomfortable.

This is a place revealed by the art of cartography. A gap in the map where there are no writhing blue threads; a place made by absence. On each side of this gap, all the burns move in different directions: anything on my side will eventually flow to the Atlantic, anything on the far side to the North Sea. Water is showing us the spine of Scotland, the dividing line between east and west.

Meanwhile the water cycle's drizzle recharges them with water, continuing their journeys in different directions. Here in the heart of the Southern Uplands, while the drizzle hides the extremities of the view, the Solway Firth and the North Sea seem unimaginably distant, as if they were terrains of different countries. Yet the water gushes forth, branching from the spine like a rib. The Annan's first steps are to fall down the hill, whether you can see it satisfactorily or not, unfailingly showing the way towards the sea. It flows through a frequently overlooked part of Scotland, a place of no great wonders or spectacles. Just fields, then the sea. The Annan, and its vanishing sources, may be a mystery still.

Millions of years after this land first formed, it became all water again. But water in its solid form. Successive ice ages over a 2-million-year period covered Scotland and most of England – then just Scotland, then just the higher ground – in a cap of ice.

Glaciers, or rivers of ice, formed in this period when snow fell faster than it melted, compressing with its accumulated weight into ice. Whereas the ice caps of the period covered the terrain – a blanket of ice that smothered the difference between mountains and valleys – the glaciers settled into the landscape. All rivers need their valleys: ice rivers are no different, reaching high into the valley heads, carving out the cwms where they come to rest at the head of a wall of a rock. Ice erodes. Ice carries with it rocks – both boulders and the finely ground particles known as rock flour – and the effect is frequently likened to the rubbing of sandpaper in some sort of act of geological DIY. Slow and weighty sandpaper on a vast scale, pushed down with unimaginable force, steady and slowly, redecorating the valley as they flowed down.

During these periods of ice – known as stadials – the landscape would have looked like that dark January afternoon by Tweed's Well. A solid white covering of ice and snow, stippled with black debris; inhospitable to life. The oldest known evidence of human life in Scotland is from the Southern Uplands: a large cache of flint tools and weapons from Howburn Farm, near Biggar; but they were dated only to 8000 BCE, four to five thousand years after the Ice Age had scoured and scrubbed and melted its way from southern Scotland.

The Southern Uplands broke free of the influence of ice early. The dramatic northwest Highlands didn't and were repeatedly rubbed by the ice into their jagged bens and crooked glens. The Southern Uplands by contrast were soothed, the effect of the glaciers more like a palette knife on rocks as soft as icing. And so the hills are steep-sided and smooth. Now small burns trickle down these ice-cut glens, dwarfed by the scale of the scenery. Since the ice, liquid water has been the most powerful form of erosion here.

Millions of years of water have made and shaped and changed these hills; Scotland's water fountain, which keeps this wet nation alive. Parts of these hills were known as 'God's treasure house' for the precious metals and minerals they contained: gold, silver and lead. But the real treasure – the most precious substance for life on Earth – is water.

There are other, contemporary, treasures in these hills and valleys: space, air and light. As you descend into the Clyde Valley from the Beef Tub, you emerge next to the six-lane roar of the A74(M) and the two tracks of the West Coast main line, snaking through this cleft in the hills. On the sides, wind farms wave from the exposed flanks of the valley above their warehouse-like operating centres and a patchy blanket of Sitka spruce and larch plantations in various states of felling. Perhaps it is appropriate that the Clyde, this river of industry, is best traced from this route, through this newly industrialised countryside: land loved for what it can do for us; what we can make from it, rather than make of it.

This ribbon of industry is a border of sorts, as I move west from the Moffat Hills to the Lowther Hills. When you come off the motorway at Elvanfoot, the Clyde branches with you. It dangles like a loose thread down a different valley to the west. It shines silver under a greying sky, looping and curling cursively through a rough, rushy field, flowing in no particular hurry. But this whole place seems in no hurry, as if time here is an idling thing. A pair of curlews hang in the air, their calls rippling like running water over the field and in through the open car window. The road is only slightly higher than the ground and the river comes and goes with the terrain.

This is the official beginning of the Clyde, as specified by the Ordnance Survey map. It happens unseen from the road.

The headwaters of the Clyde are shaped like a fishtail: two branches that join together to create the body of the river. The east tributary, the Daer Water, disappears behind its bank, tucked into the far hillside. The west tributary, the Potrail Water, is the one that appears to merge seamlessly into the course of the Clyde. A minor road turns off just beyond here, the tarmac potholed and crumbling at the edges after the harsh winter. It takes you past the farmstead called Watermeetings. I emerge from the car into silence. And then the calm is shattered by oystercatchers shrieking, a flock in the field behind me shooting off into the distance. A lapwing flaps lazy-winged over the road, a tight flock of starlings buzzes from an aerial and more curlew calls flutter through the breeze. Ahead of me is the farm's triangular field, the tongue of green water-logged land that tapers to the meeting point of these two tributaries, the beginning of the Clyde. I want to walk to it but the field is full of pregnant sheep. Beyond the flock, a pair of oystercatchers court with their heads nearly touching, their beaks facing down open, a shrill peeping duet. A lapwing is nestled into the grass as if suggesting that this might be a nice nesting spot in a few weeks' time. None of these species would thank me for jumping the gate and slogging sweatily over the soft ground. The freedom of Scottish Access laws comes with the responsibility of knowing when your presence becomes a detrimental disturbance to other lives. I leave it.

This is a paper beginning anyway, the start of the Clyde in name only. The water that flows through the river begins further back.

The Potrail follows the A road while the Daer clings to the hill edge, a crumbling minor road following it. The lane is tight. A herd of cows, split each side of the tarmac, stares implacably at each other before turning to look at me. Pines crowd round. The austere light colours the afternoon. A strange feeling descends as I travel down the road. There is no life other than a single buzzard. The moor grass is still bleached by winter. The tiny road passes shambling farmsteads and cottages with boarded-up windows. One cottage, near to the site of a lost bastle, looks well maintained but has heavy steel shutters pulled down over its windows, a twenty-first-century fortification. The river braids, widening and narrowing, flowing as if it might be unconstrained by our designs on the water.

But the Daer *is* constrained by our designs. A dam runs across it, the Daer Water, shallow and clear, running into a reservoir. It doesn't feel right. The Daer reservoir is an empty place, just grey lapping water thickening out across the valley floor; a place where a cormorant standing on a rock by its edge is noticeable for being a living thing (and suggesting unseen fish). In the drab late-winter afternoon, the Daer feels dead, the valley ahead as lifeless as the crumbling road and its boarded-up cottages, the landscape claustrophobic despite the space. This cannot be it.

A river's flow is often likened to the passage of time, an unstoppable onwards march. I have been running against the grain of the river, heading upstream against the flow. Coming at a river's source from downstream works like memory, casting back through the flow to find the origin of the event.

I need to come from the other direction; to go with the flow to find a beginning.

There is life here at Durisdeer. A red kite drifts over the skyline, a buzzard calls, a blackbird sings the joy of late March from a gable end. Durisdeer is a threshold village: perched exactly between green dale and brown glen, on the southern edge of the Lowther Hills as they erupt into a lump of steep rough ground to touch the clouds.

Dad has come to visit so I've taken him on a scenic route to find the Clyde's origins. The signpost at the village claims it is 3 kilometres long.

The calf and thigh know this track is longer than the claimed 3 kilometres. The steep-sided glen tightens as it curves east, the track relentlessly uphill without relief. Where we walk is sodden with the recent rains, water cascading down the rough stone of the path. It is a mysterious path – the Well or Wald Path – supposedly an old route of pilgrimage from Edinburgh to the Isle of Whithorn in Galloway, an early centre of Christianity. Even earlier, on the other side of the valley is a Roman road that runs past a first-century fortlet, clearly visible in the smooth stamp of ditch and raised earthwork in the grass opposite. We are aiming for the point ahead where the two old paths join at the top of the glen.

I am telling my dad all of this as we walk. He is mostly silent with the exertion. He wanted a walk and I wanted an extra pair of eyes, but I feel a tinge of guilt at what I've lured him into. We pause. A wheatear flits up onto the drystane dyke by the path. It sings a snatched phrase of song – a few jangling notes – then flits a few metres further down the dyke.

Suddenly birds are everywhere: a wren slips out of the dyke and into the heather. A stonechat – the rough-ground version of a robin – calls like two stones tapped together from the hillside. Meadow pipits jump into the air then fall with wings and tail outstretched, releasing a string of high-pitched notes as they descend. A red grouse chuckles unseen from the heather above. A kestrel breaks the skyline like an arrow shot at the nearest ridge line. A fox moth caterpillar – food for most of what we've seen – wriggles slowly across the wet path, droplets suspended as silvered orbs in its shaggy hairs. Refreshed, we move on.

The glen crests in the crook between Durisdeer Hill and Well Hill. A drystone dyke runs between the two. The Kirk Burn rises just below the wall from a boggy patch of *Sphagnum* moss and *Juncus* rush, the water flowing down the track we've walked up before falling off, down to the valley floor. Beyond the wall the waters flow north into the Clyde, the border between regions tracing out the watershed in these hills. Ahead, the bleached brown hills of Clydesdale. A curlew call echoes through the air. There is nobody else up here, just us, a sheepfold and a rusting shed, the floor peeling off and separating from the sides; a sky full of buzzards and kites and all this space.

We carry on along the track. Pilgrim and Roman; devotee and legionary walking in step across time; and now us, seeking the source of all this water. A bend in the valley opens up. The horizon is the industrial valley: a black pine patchwork on a shadowed hillside, grey turbines turning black against the sky where they slowly spin against grey layers of cloud. The pre-Clyde runs through this valley: somewhere between the dark distant valley it has collected all the braids of its burns, organising them into the two feeder flows before welding them together in that bleak field at Watermeetings.

Below runs the Potrail Water, the west branch that becomes

the Clyde. I can't see it from here but the map reveals it, slowly unravelling up the slope to this valley. It feels strange to be willing this river into life, as if my mind's eye is pushing the flow against gravity, uphill to here, long after the map sheds its name. Feeder burns trace off up hillsides. I walk down a rubble track – the rubble soggy and shining as everything is up here – towards a fork in the flow. It is my own mini Watermeetings. Two flows of clear water rippling over rocks and carving through the land, leaving this tongue, this mini proto-Clyde.

I turn back to the track that parallels the course of the small one, a 'lane', a southwest Scots word for small, slow, 'untrouted' tributaries.[5] I follow it, a dark line visible in the moor grass and *Juncus* like a hair-parting, until the top, near the border wall, where it fizzles out. I step into the quagmire, the rich peat and moss quavering under my feet. After the rains everything is wet. I find one line of water that seems more definite. It appears to flow. Or rather it doesn't look puddle-still but it doesn't ripple either. I tear a strand of dead grass into small chunks and try to drop them in to verify. They either stick to my fingers or cling to other grasses. I follow it. It disappears under a matted overhang of grass. My boots squelch. The water gurgles.

'Found it,' shouts Dad. He has stepped off the path without me noticing. While I was staring at my feet, he has been looking forward. I walk through an extensive growth of *Juncus*. By a solitary grey rock, a small triangle of water is pooling, emerald mosses bunched and breaking through the surface, the bright green grasses (the brightest thing in the valley) surviving the winter in the shelter of it. Watercress pokes through the grasses like discarded parts of a supermarket salad. Brooklime bunches, ready for the spring to shoot through with their small purple flowers. Green is the colour of water. Green are the springs where the water has flowed winter-long. Green is the life force when all else is brown.

I know what water feels like. But I am compelled to put my hand in. Not as cold as expected, not as cold as the air. I stand up; cast another look down the valley. I have felt it. I have walked it. It makes sense. The Tweed resisted me. The Annan vanished. The Daer looked dismal but here, above Durisdeer, I have found my river. A starting point for one of the flows of water that defines Scotland; that has shaped and been shaped by the nation.

I have read accounts in books and blogs about the hunt – men armed with grid references, GPS gizmos and the grit of pedantry – for the ultimate source of the Clyde, the one that extends its reach furthest from the sea, the most distant point of it. I'm not entirely sure why, because its furthest point from the sea is high overhead in the sky or deep underground. I have found one source, but water trickles out of these hills like the sweat from the pores of my skin. I can see the lanes and burns on the hills on each side that also flow into the Potrail that joins the Daer and creates the Clyde. This spring is a source. So is the next one. These hills are the cradle: the ultimate source shared by the Annan, Tweed and Clyde; a place for water that pools and bubbles and begins to flow. Here is just the first glimpse of the sky's water and the ground's water united, pulled by gravity; like my feet, that begin to flow downhill, to the Clyde.

I plant my feet. The bog stares back; a vast pile of vile jellies with a lustre like a glint in its thousand eyes, the white sky reflected in the spheres, edged by the darker water. The pile is laid in a hummock like the tussocks of rush by its edge. Not spiky like those sharp grasses but brain-like; smooth yet textured, dimpled like the impact of raindrops frozen in its surface. It would be soft to touch, of that I can be sure. I hold a trace memory of the sensation, from harmless childhood play, of its gloop; something not fully solid, not quite liquid.

I didn't expect to find frogs here, 400 metres up and nestling in the crook between hills; flourishing in the source waters. But before me lie these great, thousand-strong lumps of frogspawn, after a handful of warm days in an otherwise frozen, snow-settling beginning of spring. I don't find the adult frogs as I walk in circles through the ankle-deep water, the dark peat and dangling rushes, my eyes trained two feet in front. Sensitive to movement, the trembling peat is an early-warning system of my arrival. They don't stand their ground. But why would you when the square kilometre of the crook is all yours; the waters' source a guaranteed form of warmth, unlikely to freeze fully, unlikely to dry out. Perhaps it does take a frog's eye to see this place properly.

So while I feel the awe – the surprise that slowly makes sense, the wonder of this sign of their plenty – maybe it shouldn't have been unexpected. After all, they are as amphibious as the land below my feet, both soggy wet-solid and charged with the promise of life.

In all beginnings, another thing stirs within.

2

Actions and Reactions

Our droplet floats within its crowd, tangling itself with other droplets from other showers, flowing down other tributaries; all trickles coalescing, building flow by flow, mumbling lane to murmuring burn to talking river along the valley floor; where all rivers must seek and speak. With gravity at their back, the droplets roar around the rocks, break over falls, regroup at the bottom. Together they have power. They are the second great movement in the cycle.

The Whanganui Maori have a phrase: 'I am the river, the river is me.' It is running through my head as I stand, shivering in the black water of the Clyde, the alien sensation of a fishing rod in my hand. At shin deep I feel water against the neoprene skin of my waders, feel it tugging with each step, feel the river trying to run through me. I can see dimly the stony river floor through the sediment washing from the eroding banks. Though the water pushes, when I am stationary I feel planted, the resistance of the flow keeping me in place, as if I too am part of the riverbed. And then I lift a foot and feel its force against me, suggesting that I too should be moving downstream, like all loose things in the river.

Time passes. I have no watch but note it in the passing of birds, the growing chill inside me, the stiffening limbs. Two ospreys drift over, tracing the black line of the river, wings held bent across the wind. They are looking at the water but drift implacably down the valley, offering no indication of any fish. Things slip into a soft focus.

Unprompted, my friend Adam's thoughts have fallen along similar lines. 'I love this feeling of a thin threaded line connecting you to the river.'

I rib him, but he's right.

Earlier Adam told me that when he's been wading in the river to fish he becomes part of the river. Kingfishers have circled around him, not seeing the human. Sand martins and swallows flit low over the water in front of us. It is these still moments in a moving world that are the allure of fishing, that has driven us, Goldilocks-like, to three different places in the mid-Clyde, on a frigid April day, to find the right spot.

The river affects me, I affect the river: we may feel as though we are one with it, but even just standing in it we affect the flow, briefly, on a micro scale. The force of water is pushed up behind us, lowered downriver of us. Sediment is more likely to be deposited in front of me, though on a scale so small that it'll be washed away again by my next step. I have an impact on the river. The effect is not always so transient.

The water cycle is the basis for the science of hydrology, the movement of water around our planet, but it doesn't account for the impact humans have on the water as it moves through. The water cycle says nothing about me being here. Nothing about invasive species that we have introduced, the concrete banks, the weirs, the existence of mills, passing fishermen, the Industrial Revolution or the flow of silt from land to water. Something else is needed.

Another cycle has been hypothesised by the geographers Jamie Linton and Jessica Budds, a cycle that is aware of our influence: that water describes us and we describe and redraw the motions and patterns of water across the land. What Linton and Budds defined is the hydrosocial cycle. It is couched in the unwieldy language of important academic theory but it can be summarised simply as: there is no water on earth that is untouched by human influence.[1] Even here, where there is no obvious impact. It offers us a way of looking at water with the respect that we'd offer a human system, a way of recognising all those unintended consequences that the traditional view obscures.

It does not make an attractive diagram. It is a ring with arrows pointing in every direction. Water appears on the ring as H_2O and in the middle as 'water', the social construct created from the substance, society and infrastructure: water as a hybrid of our own creation, removed from its natural state as H_2O. Water as a substance where native species prey on invasive species that prey on native species; a feral medium of competition and control. Linton and Budds talk about it as 'perchance stabilising, perchance disrupting society'.[2]

Our waterlands have reactions to our actions.

When Adam and I arrived at the Clyde this morning we started at Elvanfoot, 5 kilometres downstream from the river's coalescing at Watermeetings. We found a dipper – a dark bird with a pale breast, built like a balled fist – standing in the flow. It shrieked once and shot off upriver as we broke the skyline and dropped down to the river's edge. Sand martins flickered, fast-winged, swooping over the river's riffles. A common sandpiper called shrilly, landing on a rock that stood just clear of the flow: it came to the kind of rest unique to its species, where its body still runs like clockwork, its tail constantly bobbing like the water, rippling fast and clear after April's overnight showers.

Adam stared intently into the tributary that joined here, looking for fish where it blurs its merger into the faster, wider flow of the Clyde, while I gazed at a condemned bridge in front of us, its green-painted ironwork arced nowhere. To the eye, at first, nothing happens here but water flowing, birds flying, a dead bridge and the distant rumble of traffic.

The fingerprints of our entwined relationship with water were everywhere. The tributary – the Elvan Water – is why the village is here, the houses running down the track to the river's edge, which in turn is why there are pines on the valleyside ahead, bare hills behind, pylons in the valley and grazed grass up to the channel's edge. Why the far bank is fixed in place.

Decisions, like water, coalesce. Decisions affect the flow of water, even here in the uplands, where the river has only just begun. The river doesn't always win.

Downriver, as we looked for a place to fish, we found a pair of crayfish claws in the bankside grass, left open as if ready to nip at a careless foot. You've heard variations of this story before. Signal crayfish were introduced from North America to English aquaculture farms in the 1970s, to help with the dwindling stocks of our original species, the white-clawed crayfish. Except it turns out that signal crayfish carry the crayfish plague, a fungal infection that they are immune to, which is lethal to the white-clawed. Think of them as aquatic grey squirrels. They began to escape. Once free, they crawled quickly through Britain's river system, part contagion, part competitor and wholly successful. Scotland shouldn't have a crayfish species at all and yet here they are, competing with the immature salmonids for food and shelter, eroding the riverbank by digging holes, supercharging the adult fish that can crack through their shells.

Having found what Adam deemed a suitable fishing spot, we settled in. Now it is mid-afternoon and we still haven't had a bite. I take a step in the river and my left heel feels sharp – cold. Suddenly. A moving sting; heel then the arch of my foot. When I borrowed these waders I was warned they might have one small hole. The sharp of the cold shoots up my nerves to my kneecap, though the water gets no higher than my shin. My right heel begins to feel wet too. One possible hole has become very definitely two, the water swilling around the feet in the rubber boots and I wonder if the trout downstream can taste the discernible tang of clammy human foot, transgressing in their element.

A man appears over the bank wearing a camouflage hat and a worried look; he is the bailiff of this beat. He wants to check our e-tickets, to have us prove that we have paid our £14 for the right to be here for the day, rod in hand, obeying the Clyde catchment area protection order of 1994.

Satisfied, he looks at my tackle and finds me wanting.

'With that rod? You're keen, aren't you? Would love to see you hook one of these brown trout. Crayfish crunchers. They grow to six, seven pound. Fight like demons.'

I don't want to tell him that this is my first time fishing since I was fourteen.

He shows us a picture of one he caught here last year: a scarlet-speckled bullion bar, burnished by the river. It is the length of the net it lies in. Its lower jaw is hooked up at the tip, a fleshy development among the trout and salmon called a kype. It forms in males just before spawning. The bigger the kype, the more dominant the male is at spawning time. This was some male.

The bailiff bids us good luck and disappears. I don't need luck. I need to be dry. In the flow of unintended consequences, our hunger for crayfish fifty years ago has led to the monster trout of the Clyde that won't go near our fake flies dangling from dainty lines. I am cold and aching from discomfort. I slosh back to land.

The river wins. I won't go fishing again.

In constructing a story that we can clearly explain and understand, the process of the water cycle has been streamlined, simplified. The reality of water's movements from sea to sky to land to sea is an intensely complicated web of directions and the luck of where the droplet lands. However, all renditions of the cycle will tell you that not much happens here, in the middle stretches of a river. That this part of the cycle is just the conveyor belt, the part where water moves en masse, from rainfall or source or the threads of its many tributaries, towards its ocean end. As if the Clyde here is merely Earth's original version of the M74 rumbling alongside.

Downriver the land getting less rugged. The brown moors become green valley. Now it flattens out. Fields of fat dairy cows instead of just sheep. The Clyde turns again, northwards. At Carstairs, in the shadow of Scotland's largest psychiatric hospital, it will take a sharp westward turn. This is a bleak, joyless place of exposed pipelines and maximum-security fencing and train tracks that I have seen many times from the windows of delayed trains to Edinburgh as they slow down to crawl around a sharp eastward bend. This is one of those rare places where we let the river make its own decisions. But it is best seen from space rather than in the place: the Clyde's westward turn is not as simple as it seems.

It helps to go back in time.

General William Roy produced the first good map of Scotland. Between 1752 and 1755, Roy compiled his map of the Scottish Lowlands, drawing cross-hatched fields onto paper, the outlines of buildings and smudged streaks for the steep outlines of hills. It was an extension to the map of the Highlands he began in 1747, to assist in the Duke of Cumberland's government-sanctioned butchering of the Highlands after the failure of the Jacobite rebellion. It's known as the Great Map – or Cumberland's map – and despite its grim, blood-soaked origins, it is the best document we have of the layout of the landscape before the Ordnance Survey was founded in 1791, the year after Roy died. It is fairly reliable. If the Clyde was doing interesting things, he would have recorded it. But his version of the

Clyde is gently meandering north, then gently meandering west after a sharp but not extraordinary left turn.

Nowadays the Ordnance Survey map view has the Clyde – in an unrealistic sky-blue colour for a day like today – curving snakelike around a slant S bend. It continues northeast of its original course before sweeping anti-clockwise around an Ω-shaped bend, a bottleneck of the river that will one day wear through at the bottom, returning the meander back to a gentle left turn that Roy mapped. When it does so the remaining three-quarters of the curve will be sealed off by shifting sediments into an oxbow lake, a boomerang-shaped body of water, thrown by the river and never returning.

This is where the Clyde's power exceeds the strength of the earth around it. This looping, fishing-line tangle of river is haunted by depressions in the earth where the channel used to be, like a pair of parentheses explaining that it wasn't always like this and it might never be again. In this stretch by Carstairs the river channel can move by 2.5 metres a year across the valley.[3] The speed can be tracked by the differences between mapping imagery: by 2023 Google's satellites showed that the original S bend has worn through halfway, leaving the bottom curve of the S as an oxbow lake, bigger than the newly joined up left–right bend. Their images reveal the old oxbows, dried up and removed from the current course of the Clyde by some way, a palimpsest in dull green fields; the water writing over land, land writing over water, erasing and making new. This part of the Clyde is an SSSI for its demonstration of the processes of a living river, a landscape following the ever-changing will of water.

Rivers change over time. They also change with time. The rest of April, May, then the first half of June disappeared in family engagements. Spring became warm, dry under dazzling sunshine. Spring is a time when light and heat and water temperature ask a question.

A field or wood answers it with flowers. A river blooms with flies: caddisflies, mayflies, stoneflies, alder flies. As larvae, the may, caddis and stonefly families eat the algae that grow where light hits water. They mutate, slough off their old skins, grow wings and break free for their adult lives lived in an eye's blink; the river light taking air, turned into food for fish and birds. Alder fly larvae hunt the other larvae, which seems unlikely as adults; winged like a Wright Brothers invention, they blunder clumsily, out of their element, into the webs of riverside spiders.

After the bloom of flies, the bankside flowers: the creased leaves of meadowsweet are topped by the whipped cream froth of their florescence. Welcome summer rain, the first for weeks, breaks against the windscreen. Adam is telling me about the time he worked here, perched in a caravan in the woods above a waterfall.

'FOC,' says Adam, like a curse. 'That's what we called Falls of Clyde. I was the peregrine ranger here for two summers. I monitored the nest but both years the egg was infertile and didn't hatch. It was frustrating, sitting around, watching an egg do nothing but it's nostalgic to be back.'

The rain passed quickly and now New Lanark is trapped under low grey clouds, stifling, thick with humidity.

The Clyde that we knew two months ago as cold and high and clear has become its opposite. We cross a lade where the water still flows fast and take our first look over a wall at the original river. To our left is the first of the waterfalls here: a knee-high ledge of rock that runs across the river. The Clyde trickles over it at its lowest point, gently slipping into a sheet of apparently still green water. To our right that water vanishes into a boulder field on the river bend. A ripple of sharply striped minnows clusters around an outflow in the shallows. Because running water makes oxygen, the fish are swimming back and forth across it, passing it through their gills, better water for breathing. To our right, in the cluster of pristine old cotton mills and tenements, a large waterwheel turns, pushed by the fast water that funnels through the lade.

It is wholly unpromising. You wouldn't look at this river today, this puddle of stressed minnows, and base your industrial utopia on its reliable power. We walk upstream to the end of the reserve.

A juvenile dipper quicksteps with the flow that remains. Its plumage is a steely grey, the pale bib creamier, the only way to tell that it belongs to this year rather than to the Clyde of summers past. A just-fledged pied wagtail sits implacably on a rock, barely moving, as if paralysed with shyness about the world. An adult grey wagtail, missing its tail, flies by in a flash of lemon yellow. In one swift movement it deposits a cropful of riverflies to its fully grown, loudly pestering juvenile. Dishevelled, it stands strangely, its motion thrown out of kilter without its tail feathers, before flitting off. The begging behaviour of young birds used to be endearing to me. Now it stresses me. I wish the wagtails well.

Beyond the birds a dog walker takes advantage of the summer-low river, leaping between the dry rocks in the river where their spaniel hops; both inching closer to the waterfalls than Adam is comfortable with. He carries with him the residual stress of coordinating the rescue of people trapped in the gorge having gone in after lost dogs. Behind us, the Bonnington hydro-power-scheme dam stretches between the banks, building the Clyde up behind us as a stagnant river. Pollution is easy to identify in the swirling off-white froth that forms a layer over the river, like the skin that congeals on cooled porridge. It's unlikely but without much warning, Drax, the company that operates the scheme and controls the river here, could open the dam. All that torpid water could wake up in a rush, sweeping all before it down the rocky steps of the Falls of Clyde.

The reserve at FOC is not the river but the deciduous woodland that flanks the gorge. It's quiet in summer but the yellow flags of cow wheat and stars of yellow pimpernel are scattered through the green undergrowth, watered by the spray that should be drifting down the gorge. Since the peregrine falcons stopped returning to their nest, nobody really visits here for the birds but for the river. It's been popular since the nineteenth century, an early stop on the tourist trail of Scotland's sublime scenery, where the path weaves its way up and down the banks of the gorge as the river twists left and right.

Before tourism – when all this was under ice – there was no river here. Instead the gradually melting glaciers pooled their water behind the edge of the ice. It created a pre-historic lake. Its outflow was blocked by ice and rubble until it could hold no longer. The ice ruptured. The lake burst its banks and the newly vigorous flow carved an alternative channel through the sandstone around Lanark until it rejoined the Clyde. The force of the flow was like a chisel cutting deep and rough to the harder bedrock that forms the waterfalls. All of this began as a chance encounter between pent-up water, hard ice and soft rock. A few months earlier, purple saxifrage, a flower of the high mountain tops, would have bloomed on the cliffs here, a relict from when all this was ice.

The river, liberated, then grew backwards. As it flowed, it eroded, cutting into the soft ground it came from. This allowed the Clyde to extend its reach up into the hills, capturing the smaller rivers that were then tributaries of the Tweed. Even the upper reaches of the Clyde, as far as where we fished in April, belonged originally to the Tweed and the North Sea, before they were rerouted by the newly powerful, post-glacial Clyde into the Atlantic. And then the Clyde flowed for thousands of years. Even a relatively new river like this shows its age. The bedrock here is marked with a series of potholes, where the eddying of water currents swept grains of sand, small rocks or large boulders round and round, grinding circular holes into the rock. One is the size of a jacuzzi, on the edge of a gentle waterfall, the water in it bubbling and swirling as it flows in then out. Others are arranged in lines, like footprints in deep snow, leading away from the river to the edge of the gorge. Each one carved by the action of water, present over millennia.

Now months of dry has stolen the water from the Clyde.

Bonnington Linn before us is supposed to be a wide waterfall, the river split around an island, falling 9 metres and rejoining below at a 90-degree turn. The water barely spills over it today.

'You always felt the spray of it, even here, through the trees,' says Adam. 'I'm shocked.'

If he thought his eyes deceived him, his senses do not. His decade-old memory of the touch of water lost to the changing climate.

In Joan Didion's 1977 essay 'Holy Water', her reverence for water is not focused on its journey through the water cycle but on its flows through the hydrosocial cycle, her awe on the machinery that delivers it. Her native California is a much clearer illustration of this than we see in Scotland. The drier the place, the keener the need.

It's in the way that the mega dams of the American West hold back their rivers into vast reservoirs, turning them into sources of power and irrigation in a dry landscape. It's in the way that the spine of California is not a mountain range but a canal flowing from north to south, its water controlled out of an office in Sacramento. If the American West is what happens when colonists attempt to subdue and tame an unruly landscape with unpredictable, extreme weather then the California State Water Project is the ultimate end goal. Water here, as Didion recorded, is a product. Orders are placed. A computer opens and closes the gates that control the flow, delivering the water down from the dams of reservoirs to the parts of California where it is needed, typically for irrigating fields. You may well have seen aerial pictures from California or other parts of the American west where the landscape resembles a patchwork quilt of green grown out of a desert. The cash crops of California's Imperial Valley, grown across 400 farms, require five times the amount of water than the population of Los Angeles would use.[4,*] The *LA Times* reported of a water official charged with stealing 25 million dollars' worth of water and selling it cheaply to farmers who consider him a Robin Hood.[5] Hydrologist Kate Ely uses the phrase 'water flows to money'.

For Didion this control over water is a 'constant meditation'; recording it in her ordered, interested sentences, laced with reflections from historic and modern California, when too much or not enough water gets in the way of life. Her interest was always in the what, rather than the why. She was a writer of place but she was not an environmental writer.[†] If this control over the vicissitudes of nature sounded utopian in the 1970s then that's because it is. Irrigation in the San Joaquin Valley flushed selenium out of the rocks and into the water, evaporation concentrated it and over a thousand birds died of selenium poisoning between 1983 and 1985.[6] California has been in a drought for twelve of the last fifteen years at the time of writing. In 2014 the drought was so bad no water deliveries were made. It is not part of the project's remit but the Colorado river has been so dammed and exploited that most years its natural end, a delta in Mexico's Baja California, no longer runs with freshwater. The delta is dead, only open to the sea, the natural movement of the water cycle severed, rerouted for our needs. If 'water flows to money' then it also flows like money: the trickle-down effect, from

---

[*] LA used 499,800 acre-feet in 2015: the Imperial Valley farms take 2.6 million acre-feet. An acre-foot is the amount of water required to flood an acre to one foot depth.

[†] I don't mean this as a criticism per se. I adore her writing.

rich to poor, is a meagre, ephemeral dribble once the wealthy have had their fill.

California is an extreme. The signs are more subtle in a wet country. Nowhere is untouched by money, though, and the Clyde is no exception. From its beginning in the bare Lowther Hills, where vegetation was cleared for grazing and not allowed to return, to the hydro-power scheme and mills of New Lanark, where the river itself is impeded to produce money in the form of electricity, to the spruce-clad valley flanks in between and the burned-heather grouse moors on the higher ground. All these things affect the river and its water in different ways because they make money. The day that the Clyde stops flowing would be apocalyptic; Glasgow and Lanarkshire recast as the Old Kingdom of Egypt when the Nile failed to flood for fifty years, the land blighted by drought, crop failure crumbling its pyramid-building civilisation. But the flow of the Clyde is shaped by a thousand economic choices. At Elvanfoot, gabions – wire cages of rocks that won't erode – keep the channel course straight and unchanging on the far-bank stretch that flanks the motorway. The river complains less at being worked on than motorists. Or rather it complains in a language we can't hear.

At Crawford, 5 kilometres further down the valley from Elvanfoot, concrete keeps the banks and bed in place for the bridges to funnel thundering trains over. Poorly sited pine trees have covered the river's gravels in a slick of silt as the rain washes earth loose from their shallow roots in the banks, changing the life of the riverbed. Overgrazing of sheep elsewhere keeps the banks bare and open and easily eroded, unlike how they would have been before the original social pressures – the first settlers of the Southern Uplands and Lanarkshire – found the Clyde, meandering at will, well vegetated and marshy, the borders of it blurrier, water and land remaking themselves at will in the soft earth of the valley. Our idea of a river now has always been subject to some form of human intervention, some change to how it flows, how the water works in the landscape, from well before any of these water cycles were even conceived of. Wherever our species settled, we have altered the landscape. The original 'natural' river is almost unimaginable, out of reach. The river pays the bill for our work with its life.

At another right turn the river falls again, a 26-metre tumble down steps of rock that would normally be unseen under the ruffles of white water, roaring and twisting again through a left turn, leaving a steep curving wall of rock, half of an amphitheatre to the power of water. The earlier rhyme, 'Annan, Tweed and Clyde / Rise a' out o' ae hill-side' continues, 'Tweed ran, Annan wan, / Clyde fell, and brak its neck owre Corra Linn.'[7] The Clyde breaking its neck was a spectacle sufficient to attract early tourists. For Victorians of a sensitive nature, less inclined to be outside in the roughness of Scotland, a building (now a ruined shell) sits above the Corra Linn waterfall. Inside it was once covered in mirrors, for revellers to feel the illusion of being below the waterfall as it thundered above them.

Illusion in the landscape is not as strange as it might sound. A century earlier, travellers looked at the scenery through Claude glasses, a tarnished mirror that would lend to the view the patina of the great master's paintings. Nowadays we look at the world adjusted through the filters of our Instagram posts, small squares onto a complex view. I find myself doing it: trying to frame a photograph where the waterfall looks impressive, when today the river barely stumbles as it passes over the falls. When the Clyde is in spate – flowing with an influx of rainwater – Corra Linn can have over 500 tonnes of water breaking over it every second.

The path descends. Just beyond a weir, the low water level has exposed a beach of grey-brown silty sand through a bank of green butterbur with leaves like rough-edged umbrellas. As usual we are not the first people to come this way. A coke can and an empty pot noodle have been discarded here; alongside dog and rat footprints left in the wet sand. A badger has run this way too. Damp sand is easy digging for earthworms. In Lincolnshire badgers digging setts in riverbanks have been blamed for their breaching, causing flooding; here the plentiful deciduous tree roots are knitting the riverbank together, regardless of what the badgers do. Something else catches Adam's eye.

He peers into the shallow water. Fish fry scatter at his shadow. He cups his hands and lowers them to the Clyde. He raises them to his face as if about to drink but lets the water run out, leaving a centimetre of tree bark in his hands. I see a flash of leg being withdrawn.

'Caddis,' he says. The larvae of the moth-like flickering flies at the water's edge. 'They build these cases out of whatever they can find in the river. There's loads here.'

Suddenly under the surface of the water I can see them too: round-edged inches, like tiny twigs, settled on the flat surface of the silty rocks. Things always look bigger underwater. The one on Adam's hand is tiny fragments of dark bark, peeled off and repurposed, tightly curving. If I hadn't known, I wouldn't have guessed there was anything living there. Out of this silty water they spin a silk that binds these cases together, half armour, half camouflage for a life on the river floor before they crawl out, pupated into a fully winged thing. Insect larvae might run the life of a river but not willingly – they have their defences.

He fishes his hands in the river again. This time he comes out bearing the smallest snail shell I have ever seen: black and whorled like a miniature thumbprint, smaller than a grain of rice.

Both the snail and the caddisfly will be feeding on the algae and the leaf detritus washed into the river from the wooded banks of the gorge. They are cleaning the river, as the falling, slowing flow of these hot days helps the algae to bloom, dirtying the water. The flow gives and takes.

While we stare at the river a yellow mayfly lands on my rucksack. Trout bait. It rises up, spinning slowly through the thickness of the air.

We are opposite New Lanark now.

Lanark is an old town; New Lanark is a new village, built in 1785 and a UNESCO world heritage site. Old Lanark is not a world heritage site and neither these days does it look very old, its history hidden by new builds, roundabouts, retail parks. Instead, New Lanark – at the bottom of a steep road that bends like the Clyde – utilised the river and in doing so shaped the industrial history of Britain. The collection of mills and tenements here, elegantly built out of dull-red sandstone, are as globally important as St Kilda, neolithic Orkney and the Forth Rail Bridge.

Its importance is down to one man: Robert Owen, a business prodigy with a sense of self-importance, who referred to his work as 'the most important experiment for the happiness of the human race'. The experiment was running a water-powered cotton mill. Like all examples of its type, the lade funnels off water from the river upstream and returns it downstream: in between, the borrowed water turns the waterwheel; a system of gears transfers the wheel's energy into the mill, where it operates the cotton-spinning machinery. Owen, who started managing mills in Manchester at the age of twenty, married the daughter of the founder of the mills at New Lanark. He then, boldly, bought his new father-in-law out of the business.

With control, the real part of the experiment began: Owen put his socialist beliefs into action, building utopia – a Scottish Jerusalem – out of his mills at the base of the Clyde gorge. He did this, in the words of Frederick Engels, 'simply by placing the people in the conditions worthy of human beings'.[8] This was unusual then. He cut the working day from thirteen hours to ten and a half, banned corporal punishment and stopped employing children, mostly orphans, to collect fallen cotton from the factory floor, under the machinery. Instead, he started a school for them. Where most mill owners saw machinery then people, Owen saw them together. His later work elsewhere, having left the mill to start a colony in Indiana, would lead to the trade union movement. But without the success of his experiment at New Lanark he would have had no takers for his ideas. And it was a success. Despite his socialist beliefs Owen was still a businessman. Mill work was still hard, long and unpleasant, and Owen was still concerned to turn a profit – which he did.

The mills of New Lanark lie sandwiched between the weirs and penstocks of two hydro-power stations. Earlier, when we walked past Bonnington power station, two large pipes ran up the hill-side, dark green and lightly graffitied, crowned with an apple-green growth of moss at the top. They thrummed with the moving of water within and muffled splashing sounds where they met the station, the sounds of water pent up and constrained. The hydro-power stations and the mills' waterwheels are two ways of utilising the power of water that seem to be in dialogue across two centuries. The mills built beside the Clyde borrow their water and return it to the river. The hydro schemes built across the Clyde impose on it their barriers and control, backing the water up and encouraging the growth of algae, weakening the spectacle of the waterfalls below, changing the river. The head knows that fossil-fuel-free energy is a vital thing; the heart knows that robbing a river of its vitality is instinctively wrong.

Several times a year, Scottish Power, the previous owner of the hydro scheme, allowed the dam to be fully opened, giving a rough approximation of what the Clyde would have looked like uninterrupted. In Highland Perthshire, the hydro scheme on the River Tummel opens its dam on summer weekends to allow white-water rafting companies to operate.* Further up that watershed, a 16-kilometre stretch of the Tummel's tributary, the River Garry, had dried up for sixty years, a victim of hydro-power before the Scottish Environmental Protection Agency began to remove a weir, rewetting the old riverbed, and allowing salmon to run upriver once again. SEPA's celebratory press releases stress minimal impact on the hydro-electric scheme.

The Scottish government sees hydro as the solution to its green-energy needs in the form of small-scale projects but this is only because all the feasible large-scale projects have already been carried out. The hydrosocial cycle still dominates.

Further west along the river is a place called Stonebyres, a salmon reserve. Fishing is banned here, although the bailiffs have recently removed 40 metres of rope used by poachers to catch the fish illicitly. It is as far up the Clyde as migratory fish get, a series of waterfalls higher than salmon and sea trout can leap presenting a natural barrier to their progress.† Those that reach here have no alternative but to spawn here, cut off from the Upper Clyde.

Spawning is something that Atlantic salmon are doing less in Scotland. Scottish government figures show a 40 per cent decline over forty years. The Atlantic Salmon Trust says 70 per cent in twenty-five years. Other species of salmon elsewhere in the world, such as the king salmon and the chinook of the Yukon River in Alaska and Canada are declining also.

---

* In the language of white-water rafting this is known as a dam-release river.

† It turns out our earlier deference was not required at all.

To be an Atlantic salmon is to be a finely honed fish, calibrated to a life that begins as an egg covered by gravel after fertilisation. After a few years as a small fish they undergo a metamorphosis that allows them to survive at sea: they drink more water, urinate less and reverse the process in the gills whereby they absorb salts from the freshwater, to expel salts from the saltwater. They head to the waters off the coast of Greenland. When it is time to breed, they navigate by smell, a sense map of home, precise enough to lead them back to the stretch of river where they were laid as an egg and covered by gravel. Spent from spawning, the vast majority die. A few survive and return over multiple years, growing to a great size. Their life cycle is marked by specialist terminology: alevin, parr, smolt, grilse, kelt. Other specificities mark their lives: they stop growing in water warmer than 22.5°C (they die at 33°C); they experience problems if the sea-surface temperature is warmer than normal in winter. At the time of writing the world's sea-surface temperature has just hit a record high.

In other countries, salmon face other challenges speeding their decline. There are 90,000 dams on American rivers, each one an interruption to the upstream swimming of a salmon and the downstream flow of sediment. Hatcheries, the industrial process by which salmon eggs are fertilised and the young born in buckets inside a building, have become a fact of life in North American rivers, despite their vast expense. The salmon they raise then reintroduce to the dammed rivers are not wild and wily, and they die sooner, unable to read the danger in the passing shadow of a heron. Jude Isabella writes damningly that 'salmon hatcheries are the sum of poor choices made over 150 years by fisheries managers, often having no idea what they didn't know about salmon, habitat and the ocean'.[9] And the hydrosocial cycle keeps spinning: the need to control and manage and make money from rivers keeps hatcheries open and offers a subsidised meal deal for the herons of American rivers.

The salmon's story of loss was once true of the Clyde. Salmon were gone from the river by the 1880s, poisoned by pollution from the riverside industries that sprang up in the suburbs of Glasgow. A hundred years later, the dirty industries had passed and then salmon showed they still had a future. They began to slip back into the Clyde, swimming up through the dark water, past factories and weirs, parks and high-rises, suburbs and grand estates. They followed their noses all the way to here.

It's good that the salmon have returned but it means nothing if they aren't capable of recovering their relationships with the river and its insect life after a sixty-year hiatus. Barry Lopez, the legendary American nature writer, has a nice way of describing this:

> I think when you're young you want to learn the names of everything. This is a beaver, this is a spring chinook, this is a rainbow trout, this is osprey, elk over there. But it's the syntax that you really are after. Anybody can develop the vocabulary. It's the relationships that are important.[10]

Lopez used that phrase in an interview he gave in 2010, arguing that rivers are living things. It's almost as if he's channelling the spirit of Owen and the realisation of the importance of the relationships that underpin everything, whether in a Victorian mill or a modern river. But this is a new way of seeing our waterways. The otherwise excellent *Rivers of Britain* by Richard and Nina Muir from 1986 cautions that 'Rivers are so vibrant and personable that it is easy to make the mistake of regarding them as living organisms.'[11] And yet I can't see a better way to view them than exactly that: a river is a living thing. The meandering, throttled, managed Clyde is a perfect example of a river that continues to live, a heart that keeps beating despite all we throw at it.

Our rivers die through a thousand cuts, those thousand tiny alterations and obstructions that we ignore because the water keeps flowing, because the river's voice is not heard over the waterfall's roar or the riffle's whisper.

Here, where the Clyde is so calm as to scarcely move and the air hangs heavy, where the mills have become a genteel day out, I wonder if Robert Owen would recognise that rivers should be a part of a union as much as workers, a thing deserving of rights and a voice to be heard. I think he would be shocked if he was alive today to see the Clyde in this state.

Two days of rain and the river has refreshed itself, the green tinge to the water flushed away. What was mud is now water. What was water is now deeper, the bottom hidden but the top few feet of water clear between the dancing clouds and dazzle of light reflecting.

Then a flash. A strange thin cloud crossing through the water. But below the surface though, below the clouds. It takes a second to sink in. Fish. Another flash. Three. Four. Six. They hang with their heads facing the flow, drifting, moving with a swish of tail, thickset oval bodies that disappear, before scudding back with the flow. My eye adjusts. It tunes into their grey bodies. Sharpens smudge into shape. Discerns their pale-tipped lips. The thin wedge of a dorsal fin. The way one nips at another, asserting itself. Size is hard to judge, especially through the distorting lens of moving water. But the smallest looks the length of my fingertips to my elbow. The biggest looks a serious catch.

A metre further down the bridge and the salmon have disappeared into the glare, hidden by the bright light.

Often the river here seems empty: just mallards and gulls catching bread, herons and otters catching the lamprey that wriggle like eels when plucked from the water. The sea trout run into the river only at night. Fish that make it have to pass the stink of a sewage works, leap up the fish ladder, slalom around the gauntlet of fishermen, chance their gravel beds lost to silt, their river lost to drought. Sometimes the salmon smell their way through regardless.

By the time I pass this way again I see the first anglers fixing flies to their lines, eyeing up the soft mud, making futile decisions, for the salmon have already slipped away.

3

The Pure, The Wholesome, The Insulted

Now the river stops roaring. The droplet canters through a well-grooved course, a lazy-looking meander as the land levels out. Suddenly the droplets meet other waters. Taste salt and sediment and chemicals; concrete and weirs and bridges and boats; currents pulling and pushing, tides running and ruffling; unusual fish, strange birds and, finally, the land receding . . .

The Clyde's descent to the sea is violent. It touches suburbia at Lanark and Larkhall, and from there on its flow shudders past the sprawl of Hamilton, Motherwell, Bothwell, Uddingston: old towns, steel works, a castle and the Tunnocks factory. Freed from the gorges, it continues to twist and turn in the old glacial moraine until it veers west at the M74/M73 junction (where it roughly follows the M74 to the heart of Glasgow, though from the road's canalised banks of concrete, you remain oblivious to this).

The final stretch of the Clyde was written by fire, water and ice. About 350 million years ago the area was all volcanic. Lava flowed, cooling into a rough, uneven rock plateau. Then, 30 million years later, this corner of the country became a mess: it was sea, then river, then swamp, then sea, river, swamp; as the Scottish landmass drifted north about 20 million years later, the climate dried and cooled; the swamp vegetation died but didn't decay, instead forming the extensive coal deposits below the city, the dubious legacy of a historical wetland.

Fire and water laid the stage. The ice came about 298 million years later, applying the finishing touches to the landscape. Glasgow sits just below the junction of the Highlands and the Lowlands; when the icy fingers of the glaciers spread out over this part of country, they moved east and south, out of the Highlands and down through the Midland Valley (the Central Belt's more geologically proper name). The glacier ground the plateau under its slow-moving weight of ice. It crumpled the surface, depositing the hillocks of clay and gravels – known as drumlins – that the city was built on. Each drumlin is shaped like a hard-boiled egg sliced in half, the thin end of the egg an arrow pointing in the direction of the glacier's flow.

This is half the story. A glacier flowed the other way out of the Highlands, turning west out to sea. The east glacier was relatively gentle. The west glacier carved and carved, gouging away at the hard rock to the west until it had cut a fjord* 160 metres deep in parts – the deepest coastal waters in Britain.

The Clyde is a post-glacial river. That is, its course wasn't settled before the glaciers but was alive to their editing of the landscape, in surface and material. After the ruptures of Falls of Clyde and Stonebyres, it meandered its way through the soft ground until just west of the city, where the old lava remains. The Clyde stays wide and shallow in its banks of old lava but at the deep westward glacial channel it changes almost instantly, becoming a sea-deep firth, held by the arms of the Highland peninsulas that hang down here.

* Fjord specifically here meaning a glacier-made narrow deep inlet of sea. Firth shares its etymology with fjord but its meaning has broadened out to include areas of estuary, bays and straits.

Even here, there's a meander of sorts: where the water of the Firth of Clyde finally meets the open sea between the tip of the Mull of Kintyre and the Rhins of Galloway, it has tracked south, curling like an elaborate hook. But this is beyond our remit: the river has become sea now. Land has no influence here.

The final portions of rivers, where the water fluctuates and changes, have always been beautiful, fascinating things, where lighter fresh and denser salty water split and flow in layers. They have also, throughout history, been treated as nature's version of a toilet: a place for dumping, where the river's flow will flush it away, cleaner water returning on the next tide.

Here, where it flows through Glasgow and out to sea, past the ruins of less enlightened industry than before, much of the Clyde's upstream simplicity disappears. The uses and abuses of it multiply. In the lowest courses of a river, it becomes our enabler – it is the gateway to the open sea, allowing trade, travel and industry; a river of arrivals and departures. The connections – and disconnections – it offers become pronounced. But in offering us its services, we give it no thanks. A heavy price is still being paid, in the visible and invisible water here; buried in the mud or carried in the body.

Because if a river is a living thing, then the possibility exists that it can die.

December. Dawn in greyscale. The east end of Glasgow. A heron stands by the Clyde's edge; the river leaden; the sun rising somewhere behind the low clouds and their smoke of rain. From a road bridge, I watch a cormorant slip below the water *like a thief into the fold*, smoothly resurfacing a few feet further across the flow. I cross the river and the road, turning my back on a building site, following a calling kingfisher into Richmond Park, but I do not see it; the morning is not split open in a blur of blue. A flock of swans swims past me towards an old man with a bag of bread. At the far corner of the park the Polmadie Burn has been fenced off.

I turn around, head back to the path between the park and the river. Under a metal bridge the fenced-off burn's water disgorges into the Clyde. It looks sludgy; concrete-coloured and sluggish, as if it might have a texture thicker than liquid to the touch, though you would never dare find that out. Two mallards dabble in it. A few bankside trees have half fallen, a suspended swoon over the water's surface. On one bough, a heron sits hunched up, head sunk into its shoulders as if depressed and regretting its choices. As the burn meets the Clyde, its water plumes, its viscous flow widening out and forming a wide grey streak along the southern side of the river and flowing seawards, almost undiluted; a river within the river.

The Polmadie Burn flows with the memory of things we might like to forget.

When the Industrial Revolution came to Glasgow, it focused on the east of the city. Hugh Macdonald, a Glasgow journalist in the mid-nineteenth century, wrote:

> If fashionable Glasgow is progressing towards the setting sun, her manufacturing industry is moving at an equally rapid rate in the opposite direction . . . where a few years since there was nothing to be seen but gardens and fields of waving grain, there is now a large community of factories and workshops, and a perfect forest of tall chimneys. The sight of such a vast expansion of our manufacturing capabilities is doubtless highly gratifying to our local pride . . . we feel but a limited degree of pleasure in lingering where our lungs are necessarily made to perform the rather disagreeable functions of a smoke-consuming apparatus.[1]

The prevailing wind would usher Glasgow's smoke away into the hills that sit northeast of the city – out of sight, so out of mind – but smoke wasn't all that was emitted here. Chemicals were produced and processed in the east end of Glasgow and anything that got into the water would wend its way through the city towards the sea.

Chromium is a useful element. A hard metal, it can be polished without tarnishing. It is the ingredient that makes steel stainless and in compound form it is bright and holds colour. It is non-toxic. But the process to break chromium free from its ore leads to it oxidising. It loses electrons and eventually forms hexavalent chromium – which is deeply toxic and carcinogenic.*

Between 1820 and 1967 J. & J. White's chemical works sat just upriver from here, nestled into a bend of the Clyde at Rutherglen. To produce bichromate of potash (now known as potassium dichromate, a chemical used in tanneries and dyeing) they would repeatedly crush chromium ore into a dust, heat it in a furnace, crush it again and then heat it once more, with different chemicals added to the mix at each change of state, reacting and dissolving out the worthless materials. Crushing created dust. Heating created fumes. Hexavalent chromium filled the air of the plant in both forms. For the workers at the plant, working twelve-hour shifts with no break, it was a death sentence in slow motion. The toxic air irritated their skin like invisible sandpaper, working away at imperfections. Cuts became ulcerated. Septums became perforated to the point where the workers became known as 'White's whistlers', each uncomfortable breath a reminder of the air that was slowly destroying them. Cancer crept through their lungs; their toxic misery ignored, unregulated until the 1950s.

---

* It is the substance at the heart of the Hollywood eco-legal thriller *Erin Brockovich*.

James White was the owner of the works. His responsibility for so much human suffering in the search for profit did not stop him from becoming Lord Overtoun in 1893 or being remembered as a noted philanthropist. What is bad for people is bad for the environment too. His firm had dumped waste hexavalent chromium across Rutherglen and Cambuslang. The effect of the waste went ignored for over a century.

Hexavalent chromium is soluble. You can bury it but when it rains – which, as two metres fall on Glasgow per year, it will do – the ground can leech polluted, carcinogenic water. The Polmadie Burn, which was fed by the West Burn that runs by the old works, apparently acquired a reputation for turning green over the years. But in 2019 it turned bright green, a lurid, newsworthy shade that no water should be. The alarm was first raised by children playing in the burn who noticed that it had changed colour. The Scottish Environmental Protection Agency (SEPA) fenced it off.

It is possible to treat ground contaminated by hexavalent chromium by injecting other chemicals in the soil, trusting the earth to carry out an intravenous chemical process. It is difficult and expensive. The solution that SEPA settled on was more primitive. They severed the West Burn from the Polmadie Burn, put it in a culvert and funnelled it straight into the Clyde, like a rogue artery pumping toxic blood into the river to flow through the city. Still, the legacy remains: in 2021, the Polmadie Burn again turned a sickly yellow; the groundwater still swills through the carcinogen-laced soil.

While the perfect forest of chimneys of the industrial age was eventually felled by the economy, the regeneration projects and smart flats that sprung up in their wake are situated just back from the course of the Polmadie Burn. And still no one can be certain about the ground beneath their feet, or what the rain might reveal from the past.

*

It was not just chromium waste; it was all waste. It wasn't just industry but a growing city surpassing the infrastructure set up for it. And no one taking sufficient care of the water.

Marion Bernstein was a Glaswegian poet of the nineteenth century who submitted to newspapers, a sort of metrical, rhyming letter writer. Her 1874 poem 'A Song of Glasgow Town' begins:

> I'll sing a song of Glasgow town,
> That stands on either side
> The river that was once so fair,
> The much insulted Clyde.
> That stream, once pure, but now so foul,
> Was never made to be
> A sewer, just to bear away
> The refuse to the sea.[2]

In the two following stanzas, she targets air pollution and social inequality, before returning to a vision of the 'crystal-flowing Clyde'. Bernstein knew that the social and environmental are keenly interlinked in a place such as Glasgow, a place where people were pushed together, breathing the same air, drinking the same water. I think her choice of word for it – 'insulted' – is perfect. The links between the social and the environment were particularly profound for nineteenth-century Glasgow.

The city was growing at an astonishing rate. In the maelstrom of nineteenth-century life – famine and clearance and pogrom and poverty – this was one of the areas that families from across Europe came to when they needed work and a bed in a new, safer place. As the historian T. M. Devine puts it, 'The rate of growth simply overwhelmed contemporary structures of sanitation and amenity in a great rising tide of humanity.'[3] That tide was an increase of 5,000 people per year in the 1820s.[4] Sandstone tenements became slums with solid walls; courtyards filled with rubbish. One room to a family. The Gorbals, in the south of the city, became the city's sickly slum, a waiting tinderbox. The river would ignite it.

For most of the nineteenth century the Clyde had been a solution and a problem. The river was the city's sewer, the outflow for the industry to the east, but the great tide of Glaswegian life needed to drink, cook and wash. The water for this was taken either from a series of wells or direct from the river, although in some cases there was no difference, the wells contaminated by the filthy groundwater weeping from the Clyde. By now the salmon on the city's coat of arms had vanished from the river, poisoned by the water.

Cholera is a bacterial disease, borne out of dirty water. It was not indigenous to Europe but as Europeans spread over the world, the sick returned home. The main symptom of a cholera infection is severe diarrhoea, in which the bacteria is still present and infectious. By 1831 cholera hit Scotland for the first time. In the city the effect was terrifying. In the first wave 10,000 Scots died, 3,166 in Glasgow. The causes of the disease remained a mystery and 3,800 Glaswegians died in the next wave in 1848.[5] The dying river was killing the people who needed it.

Various schemes were proposed to bring a 'pure and wholesome water' to the city,[6] but nothing satisfactory was found to solve the problem of the dirty water. Instead of the obvious – the unimaginable – solution of cleaning up the Clyde, the engineer John Frederick Bateman found the answer 50 kilometres to the north. Loch Katrine in the Trossachs could be dammed. A series of aqueducts and tunnels could draw water by gravity from the Southern Highlands and bring it to Glasgow. It was a life-saving achievement, opened by Queen Victoria in 1859. A monument to Victorian imagination and endeavour, it remains the source of the city's drinking water today. When cholera swept the nation again in 1866, of the 1,170 Scots killed by this outbreak, only fifty-three were Glaswegian.[7] The

Glasgow sanitary department that year had made a point of explaining the dangers of drinking dirty water.[8,*]

But until the Dalmarnock sewage works was opened in 1894, the Clyde remained the city's main sewer. It was a similar story for rivers running through other industrialised cities. In 1710 Jonathan Swift described the Thames as containing 'Sweepings from butcher's stalls, dung, guts, and blood, / Drowned puppies, stinking sprats, all drenched in mud, / Dead cats, and turnip tops'.[9] By the 1850s, the Thames smelled so putrid in a heatwave that *Punch* magazine published *The Silent Highwayman*, a cartoon in which Death, a skeleton in a black cloak, rows a boat across the Thames where festering animal bodies float; London lurking in a noxious miasma in the background. It led to Joseph Bazalgette's creation of the sewerage system, which mostly remains the same today. In Cleveland, Ohio, the Cuyahoga river was the sewer to a city of shipbuilding, steel and oil companies. The water was so frequently slicked with oil that the river caught fire twelve times between 1868 and 1969.

Some problems get fixed. Some problems change. Some problems persist. We still use our rivers as sewers,[†] still trust the water to dilute, the flow to move away. In a list compiled by BBC News of the top ten locations in Scotland for sewage discharges between 2016 and 2020, eight of the ten were directly on the Clyde and another was on a tributary.[10] Though these figures are unreliable because, appallingly, in 2021 only 4 per cent of Scottish sewage outflows were monitored. The same figure for England was 91 per cent, with an aspiration to monitor them all by the end of 2023 (though by the end of 2023 nobody could tell me if that had actually taken place).

---

[*] If cholera sounds like a Victorian-era problem, then that is true only of the Western world. The WHO estimates up to 4 million cases a year, causing 143,000 deaths. It remains prevalent across a third of world.

[†] There is more on this in Chapter 5.

We have cleared up our rivers before. We can do it again. But we need to know the problem before we can fix it.

The overnight rain was so relentless that the tannins of the riverside alder trees are washing out. A few suspended droplets of water are clinging to the bottom of the cones, their water having turned golden brown, as if the tree is weeping tears of whisky.

And then, in the grey river, a pair of dark eyes catches mine. For one long second we stare back at each other, over long noses, startled. Then it rolls forward, its thick grey body slipping into the flow of the river and it is gone. A grey seal. I do not see it again. Miracles happen. Chinks of light in dark places.

This would have been less surprising in the past. The river was once tidal here – there was a shipbuilder east of the chemical works, 20 kilometres in from where the firth narrows* – then in 1901 a tidal weir was inserted at Glasgow Green, allowing freshwater to carry on flowing out to sea while restricting the flow of tidal water upriver. The grey seal has had to hurdle the barrier, adapt to fresher water and survive the chemical flows from the contaminated land. Sometimes nature finds a way.

The Strathclyde distillery pumps steam straight into the low cloud of the day, as if the cause of all this weather, this all-consuming wet grey. At the end of the green, a grey road, a grey morning turns into afternoon: a flock of grey pigeons are eddying around a grey street tree, as if thinking about landing without committing, taking time to go nowhere like the day. The rain still falls, lightly but with a dogged insistence; it is not a surprise that Charles Macintosh, the first Westerner to create a functional waterproof coat, was from Glasgow.

---

* 20 kilometres as the crow flies, rather than as the seal swims.

Beyond the weir, the river runs chestnut, swelling and receding at the street's edge; around the many legs of the three Broomielaw bridges. Broomielaw is historic Glasgow. It's the street that held the Clyde's first quay in the seventeenth century, where paddle steamers used to depart down the river (which, due to a legal loophole that prohibited serving alcohol on Sundays on land but not on water, led to the phrase 'steaming drunk'). Now there is little evidence of anything old. I feel out of time. Redevelopment has sanitised the riverbank. In its place, bar the odd plot of empty land, are a lot of identikit steel and glass buildings. It is all very grey. The People's Palace proudly trumpets in its display of Glaswegian social history that 'the River Clyde defines Glasgow'. But the defining energy of the Glasgow I know – gallus, creative, defiantly individual and seasoned with a sense of its own exceptionalism – is not present on the waterfront.

A pair of mute swans swims close to the far bank, riding the wake from a pusher tug that sails quickly downriver. They look too graceful for this stretch of the Clyde; too white for the murk of this day. The river is widening and greying again; each ripple shifting from deep black to grey to white in the light, wave-patterned like the sea. The rain is slackening, dissolving into smirr. A raven sits 50 metres above the city on top of the Finnieston crane, built over the old Queen's Dock: a symbol of the history that wishes to be remembered. Glasgow as integral in construction and trade; Glasgow as the second city of the empire (a statement often said, the implications little thought about). An economic estuary: a place for outbound and inward flowing; transferences, morally brackish.

Over the river sits a line of brown and grey birch trees. Govan graving docks* hide behind them, disused, pending gentrification.

* A graving dock is an old name for a dry dock – to grave is to clean the bottom of a ship.

Peering through the fence that is supposed to seal the site off from the public – though plentiful fly-tipping and graffiti suggest this is ineffectual – they give a better idea of the past. Three docks: two 170 metres long, one 260 metres, each gated and with stepped sides. The only colour comes from crisp packets and lager cans strewn across the open ground and the work of spray paint. One slogan: 'Made in Shipyards'. Another, in kingfisher blue on a raised wooden plank: 'The Govan Wetlands'.

During the COP26 climate-change conference, a project was established here to grow wetland plants out of brackish water, with the grand aims of fixing carbon, providing food and regenerating biodiversity. It is a nice dream: to soften this hard edge of the city, where the land and river came together to play a role in employing so many people; that land and river could come together again to the benefit of nature, an oasis between the concrete and brick of the old city and the metal and glass of the new, a wetland in a hard place. Something to redress the balance of past wrongs. But in the two years since the project's website was last updated, I can see no signs of continued progress through the fence. Nobody answers my emails.

Instead of that dream there is now a proposal for 304 homes to be built on the site.

When I walked under the Broomielaw bridges the river was at high tide. It oscillates through 3 metres while I'm there. But the tangible effect is limited. The hard-edged river here is not given to the great changes, the waxing and waning I expect from estuarine stretches of a river. There is the sense of the maritime but only in the industry, none of the blurring of the natural worlds between fresh and salt that I expect.

The tidal weir that runs across the Clyde was designed to leave a certain height of water on the upstream side; initially for the benefit of the industry, latterly to protect the riverbank (when the weir failed in 2017, the river level plummeted and, released from the water's pressure, the tarmacked banks cracked and buckled). It has become a boundary imposed on the water, hard definition in a stretch that should be diffuse, where the change is supposed to be gradual; splitting the river in two: tidal and not; fresh and brackish, turning salty. This would once have all been estuary but now it is estuary only from this point. We have forever found fiddling with rivers to be irresistible.

It is February now. From the Inverclyde line, filtered through the grime of a Scotrail train window, I can see the river lapping at the shore metres from the track. Pale specks – goldeneye – swimming; shelducks taking off from the bank in a blur of dark stripes on white, put up by our passing. Disused jetties lie in the water like broken ribs. A magpie sits on a bush, under the arc of a rainbow that burns brightly, briefly, before dissolving into rain. The sort of stinging rain that rips at the early blackthorn blossom, shakes and shudders the freshly emerged daffodils. Within minutes it passes. I get off the train into sunshine, the warmth of light on my face, an almost nice day, the light dazzling off the wet tarmac of the town of Port Glasgow, 24 kilometres west of the city.

Here the river has changed completely, again, freed from the city's sprawl, slipping from the course carved by the gentle glacier to the deeper, wider channel. Port Glasgow is where the great transformation of Glasgow began; without this small town, the city would have occupied a different place in history, a different role in the nation.

It came about because the Clyde posed a problem for developing Glasgow. The river was too shallow for ships that could cross the oceans, fordable near Dumbuck, 16 kilometres downriver from the city (at high tide, only two feet of water covered the ford). Goods were landed at ports in Ayrshire instead and had long journeys over the moorlands. In the seventeenth century, a closer port was constructed at Newark, where the water was deeper. Newark became New Port then Port Glasgow (and the town has been confused for the city ever since).

As technology advanced, the problem of the Clyde was revisited. In the eighteenth century, the engineer John Golborne wisely decided the solution was to work with nature rather than against it. Using the power of the river's flow, he narrowed the channel with jetties of rubble – one of which, the Lang Dyke, runs for over 700 metres in the middle of the river – and the action of water scoured out the silt. Eight years after he installed the jetties, the water at Dumbuck reached a depth of 4 metres.[11] This was enough for ships to reach the heart of Glasgow. Port Glasgow became a shipbuilding town instead.

It is not long before I can see the Clyde screened by alders. A stubble of wooden posts in the blue water. This is the estuary. As the waves roll and dissipate on the shallow muddy banks, skeins of bladderwrack line the shore edge. I slip my way through the trees to stand by it. The tang of salt hangs in the air. A family of sleeping swans drifts in the bay. Two wigeon, more goldeneye, a greenshank asleep on one of the wooden posts, its steel-feathered back turned away from the light that makes its white front burn bright.

When the tide recedes, the river's freshwater flow dominates, diluting the sea's salt; on an inbound tide the salt can win. It is the mixing and movement of these waters, their dynamism, that gives them their power. In the mud being constantly covered and unveiled, flounder and plaice lurk. In summer, ospreys sit on these posts and fish for them; land birds hunting sea fish, enabled by the brackish water.

This is known as the timber ponds, and in the heyday of wooden ships, they stretched along the Clyde for a mile. In the nineteenth century, when shipbuilding was the Clyde's industry of choice, more timber was needed than could be produced here, and so ships would offload fresh-cut pine and oak. Penned in the river by the posts, the saltwater would season the wood while it waited for the sawmill and shipyard.

At the time of the timber ponds, tobacco was king. It began in the 1700s, a shady business, underpinned by smuggling and fraud, before turning legitimate. It was hugely successful: in 1758 the Clyde imported more tobacco than the whole of England.[12] The merchants involved became so rich and powerful in Scottish life, they were known as the Tobacco Lords. Then Glasgow's Sugar Aristocracy flourished, trading further south in the Caribbean; next it was cotton. Clyde-built ships sailed to Africa and America. The Clyde was vein and artery, Glasgow the heart, pumping profit from the exports and imports that the river allowed. It was a filthy trade. In the eighteenth and nineteenth centuries, tobacco, sugar and cotton weren't innocent substances but the products of slavery. Scotland hasn't always readily recognised this, as the website It Wisnae Us[13] is testament and corrective to. Glasgow's city-centre mansions are monuments to the exploitation of slaves, all built on immense human suffering, exploitation and racism. As Marlow says in Joseph Conrad's *Heart of Darkness*, while meditating on the flowing of the tidal Thames, 'This has also been one of the dark places of the earth.'[14]

The timber ponds sit in a bay formed by a small headland. It is a curious sight now, these stumps in the water, blackened and rotting; the skeletal remains of the solution to a storage problem that is hard to imagine. A more than ghostly trace of the past and its horrors.

The path skirts the headland, at times disappearing under blankets of bladderwrack, soft and stinking, and riverbank rubble, where the tide has found the lowest points. Mussel shells lie smashed on the path, plucked from the water's edge and dropped from a height by gulls, for the river's slippery protein inside. A rock pipit, dowdy in its basalt and mud plumage, picks its way between rock and weed, seeking invisible insects.

By the A8, the path drops below an embankment, running ruler-straight between four lanes of traffic and the sea defences. Squalls hit. Rainbows start small. A tight, sharp bending, I could see its complete arc in the rain, then as the squall fades and I walk, the rainbow widens, doubling, reaching the length of the visible estuary. Close to shore, a drake eider – a bulky sea duck, white-bodied with a chic black-and-green patterning – hunts for mussels too, diving and surfacing with one awkwardly in its beak, shaking its head for it to fit lengthwise down its throat. Its muscular stomach will grind the shells up. The chop on the water increases as I walk downriver, white caps cresting the waves. More squalls. The breeze increases, icy and biting, as cold as the snow-topped mountains on the far shore. Herring gulls dance in the breeze, twisting across it, plummeting then pulling up; a flight I can explain only as play.

The coastal path ends at Newark Castle, a fifteenth-century fortified sandstone house, remarkably well preserved for its weather-beaten spot on the shore. It's hard to grasp the scale of it though, for a yellow crane looms over it from behind. A ship's bridge peers over the castle's tower, as if eavesdropping. Shrink-wrapped scaffolding and warehouses stretch either side of the castle, an open-armed embrace from the industry that once hid it (for where I am standing was once the site of another shipyard). The Ferguson shipyard is all that remains of the industry that came to define the Clyde; building civilian ships out of local steel. At its peak there were 200 yards on the Clyde. By 1900, 60 per cent of the world's ships were built in Britain, most on the Clyde or the Tyne.[15]

Everything here, between the Clyde's tidal weir to the westernmost shipyard of Greenock, flowed to another tide. Not immutable, like the flow of the river, but foreseeable. If only someone had.

The world wars destroyed a lot of boats that needed replacing instantly. Immediately after World War Two, the industry stayed buoyant, accounting for 18 per cent of the world's ships* in 1947. But other shipbuilding countries had been developing, becoming efficient, modern. The global centre of the trade was shifting; the Clyde yards were confident and expanding but the orders didn't follow. By 1958 their contribution to world shipping had evaporated to 4.5 per cent.[16] Loss-making contracts were taken on until the yards began to fail, a contagion of debt sinking each one. By the 1990s the Clyde had, in Devine's words, fallen 'virtually silent'.[17]

Not entirely silent. The Ferguson shipyard is currently building two ferries for Scottish sea routes. These ferries are, at the time of writing, six years late and nearly £300 million over budget. With traditional Scottish understatement, it has become known as the 'ferry fiasco'. The ships loom out of the yard. One nearly finished, the other a little further behind; they look like artisan pieces, made with a commitment to the craft of an industry that time here has moved on from. They are each 100 metres of gently melancholy steel.

It's not the only Clyde industry to have fallen. And there is an ecological irony here. This river sustained, exported and mythologised industry. Further down the firth, it killed it. A 2010 study by two academics at the University of York compiled historical data on fish catches and mapped them against changes in fishing practice. Their conclusion was attention grabbing. The firth is in 'ecological meltdown'.[18]

Industry improves by process, an incremental evolution to greater efficiency. While the firth's fish were less affected by the pollution around Glasgow than the river's fish, they were put under different pressures. Fishing was big business, particularly herring,

* By tonnage, the measurement used for the size of a ship.

which swam in shoals that could be measured by the mile. Innovation on innovation – the seine net, the trawler, the steam-powered boat, fish-finding wires, diesel, motorised winches, echo sounders. As the catches by the old method dropped, the industry evolved another technique, a greater efficiency in finding and catching fish. When catches dropped again, they swapped target: cod, whiting, haddock, hake. Now all they can catch are the invertebrates: langoustine, crab and scallop. As the river has been cleaned up, the firth's fish stocks have not recovered. In the words of the study's authors, 'The story of the Firth of Clyde is emblematic of wide experience in world fisheries. It illustrates how opportunity and necessity have driven fisheries expansion and innovation, first to increase catches, then to sustain them even as fish populations fell . . . There can be few better examples than the Clyde of a place where fish stocks have been effectively mined out of existence.'[19]

It is easy to see the history in steel or stone. But there are also histories that we can't see with our eyes, legacies that take specialist knowledge to understand. The great poet of the building blocks of life, Primo Levi, wrote that he became a chemist to know all the things that philosophers couldn't understand. Chemistry lets us be forensic, lets us dust the river for our fingerprints. While the view of the Clyde from the coastal path shows you only the absence of the industry that used to be here, through chemistry we can read its lingering presence in the mud, invisible but indivisible from the estuary.

There is arsenic, which was used for treating the wood in timber yards and, later, for coating the metal literally to poison barnacles and algae that would otherwise 'foul' the outside of a ship. Nickel from cupronickel, an alloy that was used for ships for its resistance to corrosion in saltwater. Zinc, used for coating ships twice, first to protect timber from shipworm, then in paint to protect steel from rust because the zinc sacrificially oxidises before the steel would. Lead and chromium (again), which were also used in the shipbuilding process. Each one an insult to the river.

These metals can all be naturally occurring, but in the Clyde they are present with a frequency beyond the expected, so the chemists who studied this soberly wrote that they are 'likely affected by anthropogenic inputs'.[20]

To know our legacy definitively, we need to find other lingering substances, something completely unnatural. Something like polychlorinated biphenyls (PCBs). These are a class of man-made chemicals created for industrial uses, but in the twentieth century they began to find their way into everyday products. They were flame-resistant, cooling, insulating, made things pliable. They are also lipophilic – dissolvable or absorbable into fats – which has fewer industrial uses but does make them an excellent environmental toxin. They are carcinogenic, a neurotoxin in children and a forever chemical – although they dissolve into fats, they stay there. PCB production was banned forty years ago but they are still produced by accident, unintended byproducts in other chemical processes, and their presence is still found in unwanted places. They linger in the blubber of the world's killer whales, impeding their breeding. By the 1990s, otter faeces in the lower Clyde had nine times more PCBs than otter faeces from the upper Clyde.[21] Battery Park in Greenock was the most contaminated part of the most contaminated water in Scotland, as bad as the water in the bight between industrial New Jersey and New York. Marion Bernstein couldn't begin to know the scale of the insult we were giving the Clyde.

More squalls hit. After the shipyards, Port Glasgow has been redeveloped in the identikit way of the suburbs: retail parks and landscaped housing estates. A statue in a park commemorates the old industry: two vast shipbuilders, turned away from each other, holding hammers at the highest point of their swing. Their bodies are reminiscent of Stanley Spencer's artworks: human but gently exaggerated, the focus on their physicality. Spencer was an English war artist who painted these shipyards during World War Two. He produced images on an epic scale: 5-metre-wide panels of action, a panorama of light and dark; the clang and clamour of industrial creation; a collective of people bent over and subsumed completely in their work. The paintings were designed as predella – art to hang above an altar – as if these yards were sacred, the work worthy of veneration (he was prone to visions and raptures – he painted the town graveyard with its inhabitants re-emerging).

The statue here is finished in stainless steel, which throws the transient light of the day around; mirroring the surroundings, making the point that all of what you currently see exists because of what happened, what has been and passed.

The footpath between Port Glasgow and Greenock runs beside the river except for a long stretch, dead straight between the A8 and industrial estates and office complexes, and into the teeth of a bitter wind blowing more squalls, sideways and icy and stinging; the sort where walking becomes indistinguishable from masochism. I thought I would make it to Gourock, the town beyond, in time for the train back but time is pressing and my legs are flagging. I slump on a bench by Greenock Custom House – a grand building from 1818, in use for nearly 200 years by HMRC – and look at the river implacably flowing, blue and calmer now the squalls have passed. A black guillemot, a seabird, sits offshore gently rising and falling with the water. The wreck of the *Captayannis*, a Greek boat with a cargo of sugar, lies on its side across a sandbank in the middle of the expanse (no lives were lost when it hit another boat's anchor chains during a storm in 1974).

It is no longer river here, not really. While I had to slog along the long straight road away from the river, the OS map marked a threshold crossed. The line that read the River Clyde has stopped. The line that reads Firth of Clyde is just beginning.

What this specifically means is very little. This brackish place is where distinctions blur naturally. Definitions falter. The Clyde has never been a compliant river; it has never been fixed by a line on a map. But it feels like time to stop. I am by the Tail o' the Bank, a deep-water anchorage for ships, where land lets go of its impact on the water. The sea takes over here and the fertile seam of land and water is pushed to the edges of the firth, an afterthought.

This seems very far from the Polmadie Burn, that grey dreich dawn 30 kilometres east. It feels even further away from Falls of Clyde or the Lowthers. But this is the essence of a river. One substance, one flow: a multitude of histories, lives and deaths; things picked up by the water and its cycling. It is endlessly changing, transmuting, sometimes harnessed for its powers and sometimes feared for it; revered and exploited. The Clyde has flowed through my heart now, for better and worse, this thread of water that has flowed through beauty and bleak; bringer of life and death, somehow surviving all we've done to it.

And the raindrop has reached the sea. It will, in time, return to the air, to begin the process again. Falling as rain, somewhere, on the whim of the weather. But this part of the journey has misled. Evaporation has happened all the while our raindrop has seamlessly flowed to sea, while other raindrops cut the corner, returning straight to the estuary. Other raindrops digress. So, what happens if the clouds had pinballed another way, or never made it to the Lowther Hills? What if the cloud had tracked to the south? What if our raindrop had found itself falling and flowing into a loch instead?

Cormorant.

Bird of the black water.

Colour drains from the day. Clouds run fast. The wind whips down the concrete bank, the four lanes of traffic and the retail park; skimming the hard-edged river front. A first raindrop stings the upriver cheek.

I turn my back to the squall. A beat later and it drums into my coat, hat, hands. The wind whips the waves into the jetty in front of me, the dark river cresting white and I thought it was an otter. Present, briefly, visible as a mass disappearing, indistinct, dark; repeatedly just on the corner of vision, peripheral to my attention until – sated, bored, stoic – the cormorant emerges, calmed, riding the waves, ignoring the squall.

Across the river, these birds of the black water stand on a sandbank, half hidden by the rolling of the waves, looking like the men who walk on water, calming the surface with outstretched wings.

Milton slandered them when he wrote the devil as one, sitting on the tree of life, surveying, implacable, inscrutable; the bird of the furtive thief-like motions. Instead, they survive in the chemical hell of the Clyde's water; their broken paradise; damaged and insulted but not lost. Between the black water, the returning fish and the trees, they have all they need for life.

The squall passes. The shags at the end of the jetty sit like bottles on a wall, glowing green as the light returns. On the concrete, knot and dunlin cluster together, beaks tucked under wings, confident in numbers and grey-on-grey camouflage. The cormorant takes its moment. Two steps on the wave top and lifting off, wings unfurled, a Clyde-built bird flying further out, like so many before, just following the water.

4

Still Life

The warm air plucks our droplet. It stays suspended for days in the sky before falling, again, a breaking crown in the dark waters of a loch. Our droplet gently drifts through the water, cooling, carried by slow currents into depths it hasn't known since the sea, a chilling stillness it hasn't experienced since it was last here; all potential, waiting.

In the woods by the loch, no birds sing. Midday is cast in the sharply drawn shadows of midsummer. The air carries the faint scent of honeysuckle and is thick with willow down, drifting into the birch and beech trees and settling like an inch of dust. By the dried-up bed of a burn, there is an opening.

A gentle wind ruffles the open water of the loch. The green of *Phragmites* reeds and the full-moon flowers of northern water hemlock sway gently with the soft rustle of leaf on leaf. The sun glints off the water, brighter white than the eighty-nine mute swans that are spread across the bay, their feathers sullied by a summer of breeding. Damselflies flicker through the foreground, small charges of blue over the white crowns of the lily pads.

Lethargy reigns. From this viewpoint there is no hint of land beyond the wooded edges of this loch; no distant hills or corners of buildings to remind you of other life. It is a passing fancy but a seductive one. Imagine that this water is something self-contained and unchanging.

It is called Castle Loch. The ruined castle lurks somewhere behind the trees on the spit of land that lies ahead of me. Beyond that, the map stubbornly insists on alleging that other things exist beyond the trees: a fast road and a sewage farm and a few miles of pastoral fields descending to the River Annan. The dry bed beside me is called Vendace Burn.

In 1844, this loch and its neighbour, Mill Loch, were gaining a certain reputation for being 'plenished . . . with the celebrated vendace'.[1] Dr Johnston of the Berwickshire Field Club did not elaborate exactly what was celebrated about it – after all it neither grows to an impressive size nor is it scaled in any colour other than generic fish-silver – but he was moved to remark on its 'delicate' taste.[2] We can assume that they knew about its rarity. The vendace was found in only four bodies of water in Britain.

Not all parts of the landscape born from the actions of glaciers are the dramatically gouged valley heads of the uplands or the sheer-sided firths. Sometimes as the glaciers pushed their way to oblivion, large chunks of ice fractured off and were embedded in the ground, covered by the gravel and shards of rocks that the glaciers left in their wake. When the ice melted, the ground subsided below them, creating shallow and pure bodies of water that were ideal for this northern species, which had started moving up estuaries from salt-water to fresh. As the ice continued to retreat and Britain warmed up, this fish became stranded in four lakes: two in Dumfries and Galloway and two in Cumbria.

The first time I saw a vendace it was raining. The sort of all-soaking Scottish summer rain that colours a place and time, and tempts you into moral judgements. It was malevolent to the point where it must have been provoked; rain like a biblical retribution for the original sin of being Annan. And for several years after that, I never bothered revising my opinion of the town of rain – misbegotten, persisting, pissing rain. Annan Museum was my ark. I peeled wet coat from cold skin and carried it around, dripping on the floor. In the silence of the displays, only the thrum of falling rain could be heard. And then I came across it. A small silver fish, lying slumped on its side, pickled in a clear liquid and preserved for the future in a glass jar.

Less than a century after Dr Johnston's remark of 1844, the vendace was gone from Castle Loch. The victim, it is thought, of a poorly built sewage works upsetting the water's chemistry, enriching the water. Nitrogen and phosphorous invigorate plant growth; in water that allows the algae to bloom and dominate, suffocating and shadowing everything else. It upsets the invertebrate life of the water, which impacts the vertebrates. Mill Loch was next. Again the water eutrophied, this time due to pollutants washed by the rain from the surrounding agricultural fields. Now in Dumfries and Galloway the vendace can be found only where they have been introduced into cold-water refuges, like Loch Skeen in the hills above Moffat. They were put there in the 1990s as a safeguard for the two Cumbrian populations in Bassenthwaite and Derwentwater, wisely so as the Bassenthwaite fish disappeared in the early 2000s, leaving Derwentwater as the only natural population of Britain's rarest fish.

Lochs are places of intense details. They have an illusory aspect to them, a backdrop of life as it perhaps should be, luring you into thinking all is right. And if it wasn't for a slight smudge of green, where algae has built up around the edge of Castle Loch, you would never know this waterland was in anything other than its natural state. You would never know what it was missing; what belongs there but can now be found only drowned in formaldehyde, hidden behind glass.

It is difficult to define a lake. Where water collects and stays, stilled, where it has some degree of permanence – neither flowing immediately somewhere else, nor appearing and disappearing at the twin whims of rain and evaporation – you have a lake. Though the language is a fudge here. Stilled waters still run. Water is not good at remaining in one place, unchanging. All it does is slow down. The average time a water molecule stays in Loch Lomond is 1.9 years[3] before it either evaporates or is returned to the sea by the River Leven (whereas the water of Lake Tanganyika in the East African Rift valley stays there for 450 years).[4]

The words don't stay still either. What is a loch in Dumfries and Galloway is a lake in most of England and a lough in Ireland (but also Northumberland and Cumbria, where the word-ending is suddenly softened).* The late Brian Moss, a limnologist with an excellent turn of phrase,† diplomatically titled his book *Lakes, Loughs and Lochs*, with an apology in the foreword for omitting the Welsh, llyn. And this is just to mention the examples beginning with 'l'. Even more wisely, Moss knows that trying to fix a definition on water is dubious business. He writes:

> Think of the continuous thread of water molecules as they move from the raindrop to the middles of oceans and are cycled back by evaporation, then rain and snow. Those who classify habitats . . . have hindered nature conservation and wise management of our landscape. Such classification emphasises boundaries and separation; it is the affliction of living in a land divided by human activity for millennia.[5]

So, sensibly, he begins his study of lakes (or lochs, or loughs, and so on) by defining them as places where the water that flows in is not

---

* And you will know me as an Englishman in Scotland by my incorrect pronunciation of 'loch'.

† Someone who studies lakes.

replaced within a week. The distinction between a lake and a pond is subjective – a matter of scale that you just intrinsically know, though the Scottish term 'lochan' blurs this boundary: a diminished lake, an exalted pond. Moss's approach seems a fair starting point, because if we trace his 'continuous thread of water molecules' to where water collects and slows down, we find ourselves introduced to unfamiliar life, a complex chemical balancing act and an environmental archive. All that a river would have washed away is expressed within a lake. Water is the threshold to geography: it tells us about places and their past.

Water collects mostly because of glaciers. When these rivers of ice rumbled through the landscape, they did not work evenly. Their erosion gouged and scraped and rubbed. In places they stripped the landscape back to its hardest rock along the valley bottom (the most common type of lake). In others they deposited gravels (or parts of their icy selves splintered off, as at Castle and Mill Lochs). Sometimes the gravels they left can block rivers, such as at Slapton Ley in Devon, where the River Gara was blocked from draining into the sea by a large bar of shingle, the backed-up water forming the largest natural lake in southwest England. By roughing up the land the glaciers allowed pockets of water to stay, to be held. Elsewhere in the world, where the tectonic plates pull apart, they create deep gaps for water to flow into, such as the 1.6-kilometre-deep Lake Baikal in Siberian Russia. Sometimes the movement of rivers can throw off parts of themselves into lakes. Sometimes we built them, either to collect drinking water or for recreation. Others, like the Norfolk Broads, were made unintentionally, when old peat diggings were flooded.

I think of a lake as being a leaky petri dish, with water as its medium.* They are containers of life – fish, insects, birds, plants,

* Moss writes that it is 'hard to take an inspiring photograph of a lake, or at least what is thought of as a lake: the mass of water'. I appreciate that my analogy is also not the most inspiring but bear with me.

freshwater planktons – and while the water molecules continue their cycling, the life generally stays put, growing in the confines.* Even the word 'lake' finds its origins contained in the Latin *lacus*, meaning a basin.

As we are a meddlesome species, we perform unwitting experiments on these petri dishes. We pollute; we alter the chemistry of the water; we introduce and extirpate the species that live there; we poison the well. Only 14 per cent of English lakes currently reach the threshold for good ecological status.[6] But the petri dish has an outflow and its levels of water and our inputs change over time – for good and bad.

The struggle for specific definitions is one of the limnologist's burdens. Another is the vast gap between the simplicity of what a lake seems to be and what it actually is: of all the habitat-focused fields of ecological study, limnology is possibly the most opaque to outsiders. We see water, plants, wildlife; they study the parts we can't see – the chemistry, the minuscule life, the history, the way the water behaves in the lake and so on, into progressively more obscure niches, where the language is alien and the arguments remind me of Byron's great putdown: 'I wish he would explain his explanation.'[7] Not all of it seems necessary to the layperson.

But there are parts of limnology that are useful. It has methods for measuring and classifying lakes, such as the trophic state index, by which lakes are sorted according to the nutrient levels – which plants and algae need for growth – in the water, an approach that underpins how conservationists see lakes. The index defines the richness of water and how easy it is for aquatic plants and algae to grow

* If the outflow is right, some migratory fish will enter and leave the petri dish.

via, as Levi puts it, the 'solemn poetry, known only to chemists, of chlorophyll photosynthesis'.[8] It ranges from oligotrophic through mesotrophic to eutrophic; or from low biological productivity through moderate to high. Some also offer dystrophic as a category, for extremely acidic, dark-stained waterbodies (see the dubh lochans of Chapter 6), and others include hypereutrophic for bodies that go beyond the normal categories of productivity.

It is a neutral scale: it offers no comment on the water quality, for what is good quality for a dabbling duck is not good quality for a vendace or an Arctic charr; the water a swimmer or paddleboarder might think of as good is not the same as how an entomologist, angler or botanist would see it.

There is a nutrient paradox involved, across all waterbodies. The more nutrient rich the water is, the easier it is for life and the worse it is for life. As nutrients seep into the water, they boost the plant life that is particularly eager for them. The algae, charged by the nutrients, spread like a carpet across the water surface. They become greedy for light, shading over the other aquatic plants and starving them of the sun. Unable to photosynthesise, they then die. Death uses oxygen: the bacteria that breaks down the dead plant matter respire, further depleting the oxygen in the water. If unchecked it continues until the fish choke, the invertebrates die and the water is deemed lifeless (although full of algae, still living).

These trophic levels are a natural process but each can be altered; made more extreme or diluted by our activity. Some have seen these trophic levels as a way of telling the age of a lake, assuming that as a lake gets older it inevitably, irretrievably, becomes eutrophic, as it fills with sediment and gradually enrichens with nutrients (although others critique the assumptions behind this). Whether this is accurate or not, it points towards a truth that is hard to see from our perspective, when the existence of waterbodies seem so definite and present and immutable. The truth is that all lakes are transient and temporary; on a geological timescale they are a blip, ready to fill up with sediment and plant life and turn into bog or fen; from there woodland can grow and the wetland vanish, discernible only as a contour line on a map aeons into the future, or an isolated deposit of silty or peaty soil.

Water is the medium of the lake as petri dish. But water is not a simple substance; it contains multitudes. Where it collects, it allows communities of plankton to develop.

Plankter is an ugly word. It is not often used for it refers to a single organism of plankton, organisms best dealt with in the plural. Brian Moss, visualising what an osprey sees of the water as it drifts over a lake, is dismissive. 'It's all a matter of scale,' he says, warming up to a damning indictment, 'Osprey and human see only gross features. A two-millimetre planktonic copepod faces a very different, viscous world of great intricacy and peril.'[9]

The problem is that to view water in detail, to see it deeper than the gross level, requires a microscope instead of the sharp eye of an osprey. A 2-millimetre feathery-antennaed shrimp is hard to see at all. The maths is ludicrous. Moss cites that an 'exceptionally dense zooplankton community might have a fresh biomass of one gram per litre, and occupy perhaps a tenth of one per cent of the water volume'.[10] It gets more mind boggling: in a 'typical' loch of 100 hectares, Moss calculates that there would be 30 million tonnes of water, 15 tonnes of phytoplankton and only three tonnes of fish.[11] And it's the phytoplankton and the zooplankton – in concert with the chemistry of the water, light and temperature – that make the life of a loch work.

It gets more mind scrambling as you delve into all the different types of plankton to be found here. Cyanobacteria – which you may know by their old name, blue-green algae – are a division of plankton found wherever there is water. They are one of the oldest life forms on the planet, older than oxygen, which they evolved to produce through photosynthesis.* These bacteria are known as bacterioplanktons. Virioplanktons (viruses) are also commonly found in water. Their specific roles aren't entirely clear. As Moss says, 'Plankton are the simplest community . . . but simplest does not mean simple.'[12] There is even the 'the paradox of plankton': plankton exist on a more diverse scale and scope than it would seem the water can provide for.

All phytoplankton want to do is turn light into energy and oxygen. But in a lake they have to deal with the fact that they are always slowly sinking. The deeper they go, the less light they have, the less they can photosynthesise, until they settle, dying among the sediment and rocks of the lake floor. Unless, of course, a wind catches the water, swirling an eddy or other current, drifting them back up to the light-filled surface of the water, where they can photosynthesise again. But water is viscous. Sticky and tricky to drift through. And the zooplankton are trying to eat them. Phytoplankton life is a compromise between all of this: each species is a different solution to the same basic problems, different shapes and colours and sizes. Looking at phytoplankton under a microscope is like seeing one part of a kaleidoscope. There are strange shapes: some species look like stars, some like mouths screaming and others like hairs or lichens, rings and circles, as if they held more in common with the shapes of deep space than shallow sunlit water.

The zooplankton has more agency and even more variety. As well as the copepod shrimps, there are flatworms, Ostracod shrimps, which look like a sprouting seed, and water fleas – *Daphnia*, like a badly drawn stag's head, and *Bosmina*, which through a microscope

---

* Leading to an extinction crisis in early life that wasn't equipped to deal with oxygen in the air.

resembles a legless elephant. And then there are the zooplankton that are harder to imagine, species we run out of useful language for: rotifers (they look like an insect as imagined by Francis Bacon) and ciliates (a protozoa; they are essentially just a blob); things obscure but capable of moving and eating as they drift through the water. And, in turn, they too are eaten. The minuscule ciliates are eaten by fish fry. As fish develop they eat rotifers and copepods before turning to more substantial food sources. For a fish such as an Arctic charr, a population in a lake can split along the lines of diet: those that eat other fish grow the biggest and live the longest; those that have a diet of plankton become the brightest, their breast scales a searing scarlet, before they flare out.

The current answer to the paradox of the plankton is that a lake is essentially in a state of chaos at that planktonic scale. A revolving cycle of boom and bust. No one type of plankton should come to dominate for long enough to drive the others to extinction, as if evolution hedged its bet, to ensure a diverse population of plankton. On this scale that we can't see, lake water holds hypnotic strangeness; creatures that defy comparison and render maths incomprehensible, unwieldy numbers that slip from the mind's grasp. Water is the medium of the loch but it is the complex cocktail of plankton that runs it. If kept in check, they enable all other life; if the chemistry tips too far, they take over the eutrophied lake, like unruly prisoners evading their guards, suffocating everything else.

It's a natural process. But we inevitably alter it, charging up the changes. Because we have wanted forever to be near water. The urge drew nearly 20 million tourists to the Lake District in 2019,[13] where England's largest lakes sprawl along the hard-bottomed valley floors, and busy villages cluster along their edges (and where most of my childhood holiday memories are from). Nowadays the weight of visitors can overwhelm. In Cumbria those towns are served by a leaky sewage treatment works, which has contributed to the eutrophication of Windermere, jeopardising its population of Arctic charr.

The Arctic charr is a beautiful fish, a salmonid with a dawning pink-peachy blush along its breast; the most northerly freshwater fish in the world and the first to return when the Ice Age receded, leaving cold, nutrient-poor water in its wake. If we can't fix our relationship with water, it seems as though the Arctic charr will go the way of the vendace, lost from all but the remotest, deepest, coldest, cleanest lakes.

Windermere might end up like Lough Neagh in Northern Ireland, the UK's largest lake,* a sickly green in summer; its algae blooming from the nutrients that slide off the fields into the lough's tributaries during summer rain. Lough Neagh is one of the few waterbodies of note in the UK where the term hypereutrophic applies; where protestors in the autumn of 2023 held a wake with a coffin, in mourning for what they saw as the passing of the lough.[14]

A breeze ripples Loch Stroan, the sunlight casting white rings on the loch floor. The water changes colour with the light. In the shaft of sunlight it turns amber, the light illuminating the slight stain of peat that gives a temporary tan to my hand thrust in the water. Fry swim over the fine gravel on the loch bed that looks like a golden beach. A migrant hawker dragonfly zips, unzips, zips up the air back and forth along the overhanging bushes. The light goes: the water suddenly opaque with drifting cloud.

I am left on shore holding the toddler. Miranda and a friend wade slowly into the loch, neoprene shoes on sharp rocks. Water

---

* With an area of 391 square kilometres, compared to Loch Lomond's 71 square kilometres. It still contains less water than Loch Ness.

turns from lukewarm to cold. They slow down even further. Crouch low as the chill creeps up them, lower still until it looks as though they are sitting in water, then suddenly they are floating.

In Dumfries and Galloway, the water is eutrophying in the east, in the richer lowlands around Castle Loch where the human pressure is more tangible, visible in its phytoplankton. Head west, away from the towns, and the lochs progressively turn oligotrophic – low in nutrients, high in oxygen – as altitude increases, as acidic granite intrudes into the geology.

This loch is oligotrophic – stony-shored and beautiful, as they tend to be. There is little algae here: the light passes through water that looks dark but is clear, without the nutrients necessary to feed much plant life. Here, in the forest's valley floor, the water remains clear, cold and deeper (12.5 metres) than it looks. Without algae colouring the water and decaying, it retains the oxygen that is given to the loch by the wind rippling the surface and the falls that the inflow – the Black Water of Dee – crashes over. Oligotrophic water is defined by what it doesn't have: a jigsaw of absences that creates the picture of its life: home to the hardier, those that can't handle where the living is easier.

At the shallow end, near where the water flows out of the loch, there are small stands of rushes and patches of lily pads, a meagre growth of greenery. A red kite drifts over and it will surely see the loch as a blue-black puddle of sky in the green carpet of forest, the small rings of water where the lips of fish rise to the insects trapped on the water's surface, the heads of Miranda and her friend and not a lot else. Except for me, holding a complaining toddler who is desperate to make her own decisions, such as getting in the water, regardless of the consequences. We reach a compromise: her up to her knees, the water filling her wellies, soaking into the socks that we thought would be a good idea to leave on. She sits on a rock. She kicks her feet. She feels for the sticks and stones of the loch bed and looks at her reflection as it appears and disappears with the passing sun.

Miranda drifts back and helps the toddler wade in further than my feet will let me walk. I stay crouched. I normally look at lochs from six feet up but I like this new perspective. At her height the water seems vaster, the ripples faster. It shrinks the trees ringing the edge. Everything seems bluer, more compelling.

Later, as we sit in the car, toddler fast asleep, I wonder what I lose by not swimming.

'The feeling of cold, outdoor, non-chlorinated water is amazing,' Miranda says. 'The cold is addictive. Sea water is good, but loch water is like silk. It has a depth and a darkness and a thickness.'

'A thickness?'

'Yeah. It's like the difference between drinking your parents' tap water [Suffolk, hard] and ours [Dumfries, soft]. And it smells of iron and freshness rather than the weird medical smell of a pool.'

She pauses. Then:

'Swimming is like flying, in a way, to not be touching anything solid with your limbs. There's freedom in looking across the water's surface, knowing that you're moving away from land, from where you're meant to be, and that you could keep on going. In the sea that feeling is almost limitless. But in a loch there's a sense of joy that you can swim towards its centre.'

Sensation and centring. I am less bothered by needing to be at the centre of things. But the sensation piques me. I think sensation is what I miss by not swimming. I get the sense of being untethered by gravity, the amniotic sensation of 'how it was before you were born', in Roger Deakin's phrase (though as our midwife gleefully pointed out, amniotic fluid is mostly foetal urine, so I can think of nicer substances to be untethered in).[15] I have, occasionally, been roped into paddling a canoe and it feels halfway between flying and walking, like gliding to an iambic beat; while the motion of water pushes and pulls through a secondary skin of polyethylene. As close to swimming as I will get.

The science behind how we feel water – the sensation of wetness on skin – is a curious one. Human skin does not have a hygroceptor, the sensory receptor for wetness (whereas insects and arachnids do). Instead, it is currently understood by neurophysiology as a sense that is learned through touching a combination of temperature and motion. The most important factor is temperature. Testing with the same amount of wet stimuli at neutral (30°C) and warm (35°C) temperatures were regarded as being significantly less wet than cold (25°C) temperatures. Tests with a moving stimulus were regarded as wetter than stationary.[16] It might be that cold-water swimming feels significantly wetter than in a pool just because of the temperature and the natural movement of the water within the lake. Something I will, barring an unexpected change of heart or a freak accident, never actually feel.

Between the trophic extremes – water either bottle green or window clear – comes the rarest thing: the middle ground. Mesotrophic is a balancing act of nitrogen, phosphorous and calcium carbonate; a loch whose water is neither particularly acidic nor alkaline.

Mesotrophic lochs are hard to find. As fits with their status as the middle ground, they are mostly found on the threshold of the uplands, where there should be less impact from agriculture and tourism, less nutrients and effluents seeping into the water. Their diversity of plants is thus good: the water is balanced with enough nutrients to stimulate growth but not so much that they encourage the phytoplankton to cloud the water, still allowing light to penetrate through the depths. This in turn benefits invertebrates, specifically the caddisflies, for whom mesotrophic waters are their most valuable breeding grounds; the right amount of clarity and life.

We pull up at Woodhall Loch early. It sits in the rough between the open ground and the hills, bright but cold below a clear sky. Adam has come in fishing mode: polarising sunglasses on, he wades into the water, parting the thin screen of *Phragmites*. 'Slippery,' he says. I have not come in fishing mode and I hang back on the shore, under a birch, peering at the clear water and its bed of silt-slicked stones. Reeds are unusual on the Galloway lochs but here they are scattered around the loch edge, not in the dense screens I am familiar with, but a thin growth, water and light visible through their stems. Marginal here but thickening; we can see the underwater growth, the off-pink sheath of a reed rhizome rolled up and pushing up through the stones. There are sparse starbursts of white water lilies and a single swan swims through the dark water into the brilliant blue of the reflected sky. At the top of the loch, rings part the water's surface; fish that we can't see, only discern from their effect on the water, rising for flies that we also can't see, caught in the surface tension.

These lochs are difficult to read. Possessed of a subtlety that can be hard to see as they are neither one nor the other but tread the fine line between the two. This one is doubly so, for it was historically oligotrophic. But, surrounded by pine plantations on one bank and areas of cleared ground where cattle graze to the water's edge, it has become gradually enriched with nutrients, mutating into mesotrophic. That slippery silt is one hint; the creeping rhizome of the reeds another sign of life moving in, of the water becoming enriched.

The forecast for here is not particularly positive. It seems inevitable, without effort, that the nutrient levels will continue to rise, phytoplankton will respond and the loch will carry on greening to eutrophic. There is no plan that I can find to protect the water from what is happening to it, despite mesotrophic being the rarest type of lake – and getting rarer. Meanwhile a campsite has been opened and fishing is newly advertised; the infrastructure of attractions, to bring people to the banks of the loch and keep them here. The other local lochs that were mesotrophic are already becoming eutrophied. Trophic state index is a bit of a blunt-force tool. Water is slipperier, more fluid than its definitions, as this loch, mutating, moving between states, shows us.

If a lake is a petri dish where the water is ever changing, we need to consider the dish itself. This is where the analogy falters: Perspex does not change. The sediments containing a lake do. They leave us a record of change that we can read. They tell us what has happened in the past and what is happening now.

This begins with rain.

Robert Angus Smith was a curious man. In his work as a chemist, he had wanted to test the air for its chemistry but, lacking the proper apparatus, he settled for testing precipitation, suggesting, 'If there is life and death in the air, we must believe the same of the rain.'[17] He chose to drink the rain. His 1872 book *Air and Rain* records some of his tasting notes. The rain from Greenheys, then a suburb of South Manchester, he recorded as: 'when first got, very greasy; after standing a while this taste becomes bitter, like rotten leaves . . . sickly taste begins when the greasy and bitter tastes are gone'.[18] And so, by accident, he discovered acid rain.

Smith prefaces his revelation with the simpler admission that rain is 'complicated'. This sums up his writing style too. *Air and Rain* reads like highly educated homework, a step-by-step setting out of things he did and thought in the order that he did and thought them. In parentheses he disagrees with himself. It is a slog to read, fogged down by his inability to summarise, bogged down by the weight of observation and statistics. Then with a startling clarity he'll state that 'all the rain was found to contain sulphuric acid in proportion as it approached the town', adding, 'if it is acid, artificial circumstances must be suspected'.[19] He suspects correctly that the industrial burning of coal is to blame.

Smith did everything right. He made the discovery, identified the cause and wrote about it. The curious thing is that nothing happened. It might be the tedious style of his writing, or apathy towards the effects he listed – crumbling stonework, rusting metals, fading paint, rotting wood, Glasgow's mortality – but the term fell immediately out of use. The Alkali Acts* of the late nineteenth century helped reduce some of the most toxic pollutants emitted (and Smith later gained employment as an inspector, policing these emissions) but the coal kept burning, rain kept biting into stonework and the UK government didn't pass a Clean Air Act until 1956. And even that was too late, four years after the Great Smog of 1952, which blanketed London with a stationary fug of acid rain for four days, causing 12,000 excess deaths.[20]

Rain is a cleaner of the air. Pollutants, in this case sulphur dioxide and nitrogen oxides released by the burning of coal and the driving of cars, react with water, turning into sulphuric acid and nitric acid. Water droplets absorb them, then fall to Earth with their chemical interest. Purifying the air, polluting the land. What that acid rain falls on is important too. Limestone and chalk are alkaline and can neutralise the acid as the water moves through them. Granite and peat are naturally acidic; the rain passes through them unchanged. The soil doesn't stay unchanged though: at 5.5 pH aluminium becomes soluble, leeching toxic metal into the soil, poisoning plants further, tainting the water.

*

---

* The Alkali Acts were targeted at the alkali industry. Their emissions were acidic.

My friend Charles is known as Trees. It is a work nickname, for he is the only forester in his land-management office, but it holds true on the weekends. He is behind the wheel of a Subaru Forester in tasteful green, his windscreen wipers frantically opening brief windows of vision as the rain blurs and obscures. At the western end of Loch Trool he parks up. He dresses as a tree too: brown boots, black trousers, jacket the green and blue of Sitka spruce. Hood pulled over baseball cap. Green waterproof cover over his rucksack. Trees possesses a walk that he says all foresters have: a half-sprint on a straight line, over or under or through any obstructions, including but not limited to hawthorn, blackthorn, bogs and clear-felled hillsides. He goes and does not stop, and I have to keep up.

Rain closes in around us. Soaked in minutes. Green coats washed slick against forearms. The hat I wear instead of a hood sodden and useless for the day.

The path we take weaves through mixed woodland by the grounds of a farmhouse where the legendary conservationist Derek Ratcliffe stayed on his first visit to Galloway, a solo exploration as a sixteen-year-old, searching for peregrine nests in the hills. (Sweetly his memoir records that his parents made him swear to 'take great care'. He did not.[21]) He would later make his name tracing the effects of chemicals on peregrine eggs.

We crest a small bluff and as the path zigzags down, Loch Trool appears again to the side as the view opens up. By now the curtains of rain have drawn back as far as the hillside, catching on the exposed rocks and fraying, settling in the gaps between peaks like old cobwebs in the corner of a room. The loch lightens as the sky does. As the scenery opens and closes, so does the flora: ling and bell heather in pink and purple; rowan, ash, alder and hazel; breaks of pine plantations and clear fell; young conifers and some left imperiously old, the tallest living things in the landscape. Their bark mapped with contours of lichens, lined like the ripples of the pale loch shining in the background. You would be forgiven for thinking that this view was incredibly Scottish.

Nestling in the crook of a narrow glen, it is difficult to get a view of the loch in its entirety: something is always getting in the way, the trees always hide something of the view. It makes Loch Trool beautiful. It makes Loch Trool popular. It's why at the far end we take a joining path up into the hills, connecting with a forestry track, to a loch less travelled.

I look back at the far end of the loch. After the rain, drifting clouds of evaporation curl their way through the pine tops like tendrils of smoke; the briefly visible afterlife of rain before it vanishes, leaving only a dampness. Raindrops may seem transient to us but not to the lochs. When John Keats died, melancholic about the impact of his poetry, he insisted his gravestone read 'Here lies one whose name was writ in water' to reflect that passing transience. But what the rain writes is a different matter. Sometimes what is writ in water doesn't disappear but gets filed away in the archive of sediment.

Acid rain is an uneven problem. The same currents that move water vapour around the atmosphere are what move pollutants around too. Here in Galloway we have acidic peat over, in part, acidic granite. So when acid rain falls onto an already acidic surface, there are no buffers. The acid is free to flow into the watercourse. The effects did not take long.

By the late nineteenth century, at Loch Enoch – a short distance on the map from here but an awful walk, via either Murder Hole or Dungeon Hill – deformed trout were beginning to be caught. Specimens were collected, exhibited, speculated on, treated as curios then stored in museums. Fins were warped. Tails shrunken, like a fingernail bitten to the quick. The term 'dock-tailed trout' was coined.[22] It wasn't long after these trout became famous that they disappeared, gone from the loch by 1890. Then the lochs Neldricken, Narroch and Valley lost their trout too, loch by loch; the same pattern as at Enoch, deformity then extinction. It was seen on Islay and near industrial areas too.

Malachy Tallack makes an important point in his book *Illuminated By Water*:

> The existence of trout relies upon the health of the water, the health of an entire ecosystem. It relies upon an abundance of insects, which in turn relies upon and sustains an immense network of other living things, of microorganisms, plants, birds, amphibians. Trout are proof of life, in other words. They are the vitality of a habitat, embodied.[23]

But this wasn't understood in the nineteenth century. Water pollution was proposed but rejected; Galloway and Islay couldn't possibly be polluted, with one Gallovidian stating in 1927 that Loch Enoch is 'away from every possible source of contamination'.[24] If Smith hadn't been ignored, how different things could have been. By the 1980s it was known that fish fry are deeply susceptible to water quality affecting their development. We also knew by then that the diatom record showed that the lochs had been acidifying since the nineteenth century.

A diatom is a bit harder to define than a trout. It is one of the phytoplanktons, with a skeleton made of silica that encases a single-celled body containing chlorophyll – though to add a further layer of clarity or confusion, they don't appear green but a golden colour when observed alive through a microscope. They photosynthesise, contributing an amount of oxygen into the atmosphere. That figure is contested but when pushed to make a guess the paleolimnologist* Dr Jeffery Stone reckons it to be about 'one in every four breaths'.[25]

* Someone who studies pre-historic lakes.

Diatoms are found in every wetland. They are of interest because when they die their glassy skeletons sink into the sediment below the wetland they live in, forming a record we can read today across millions of years of what lived where and when, even where the lake no longer exists: the dust from the Bodélé depression in Saharan Chad that blows across the Atlantic to fertilise the Amazon rainforest is diatom dust (left from the period when Lake Chad was nearly twice the size of Great Britain*). Diatom species are finely calibrated to their environment. They change in response to many things. Forensic scientists can use their presence or absence to determine drownings and the locations of crimes.

At Loch Dee, nestling in the hills above Glen Trool, they are used as an acid test. The loch water is regularly sampled as part of long-running studies into the health of the lochs of Galloway. The samples, analysed, show the diatoms present. Cores of the loch sediment show the change in diatom species, datable to when they changed. Other elements back that up, as does a pH level. When the lochs of Galloway were found to have been acidified it caused an argument: after World War Two, the Forestry Commission (now the devolved Forestry Land and Estates Scotland) bought up the hard hill farms of a huge swathe of Galloway and turned them into plantation pine forests.† This change in land use could have caused the acidification, the Central Electricity Generating Board argued, pitting two limbs of the government against each other.

The CEGB funded the science. Limnologists Rick Battarbee and Roger Flower's analysis found that actually the lochs had acidified long before Galloway was forested, beginning not long after Smith discovered acid rain. But it was a pyrrhic victory for the Forestry Commission. What Battarbee and Flower also found was that the

* It is currently a little larger than London.

† Read *Native* by Patrick Laurie to understand what was also lost in this.

forests were definitely contributing to the acidification: they called it 'the forest effect'.[26] The shape of pine needles encourages particles to be drawn from the air. The low cloud that frequently bathes the forest top essentially bathes the pines in polluted air. When the needles fall, they usher these acidic pollutants to the water course.

The Loch Dee project was formed.

It is a long slog round the corner of the hillside, high above a burn running clear. Between us and the burn is a sloping field that was forested but now bears the scars of clear-felling. Furrows cut into black peat, deep with water, tinged green with algae.

'It's not particularly good,' Trees says. 'It looks horrible. If this now gets disturbed, if you get a flash flood, if the furrows break, it can all just go down into the watercourse.' And it would take with it its load of acids and nutrients; pollutants scavenged and stored, ready to be released.

I like talking with people who work on or with the land. They are aware of the moral complexity of it. Trees talks lovingly about his forests before he supervises their felling. Another friend, a forest ranger, looks up at a larch with affection and then talks of it as being a lovely product, though *Phytophora*, a fungal pathogen, is turning the product in this forest to dead grey ghosts. We all love nature. We all want a better environment. We just disagree about how to achieve that, and remain friends.

The clouds lift, releasing the peaks. The sky turns blue in patches, as if reluctant to commit fully to the end of summer. We are above the bulk of the plantation that we can see. Moor grasses are turning brown. Living and dead bog asphodels stand side by side. A white devil's-bit scabious stands out like a snowball between the regular lilac pom-pom flowers. Eyebrights gaze up between the stones of the track. A Scotch argus flutters past on end-of-season wings, ragged and weak. Pond skaters skitter down the drainage ditch by the track. We crest the top of the short ridge and see the path bend down below us, running to the side of our destination: Loch Dee, nestling in the palm of the hills, at the head of three valleys and the source of the Black Water of Dee. In the new sunshine it looks perfect, wild, natural.

Rosebay willowherb forms a thick pink stand by the water's edge, running between birch, rowan and willow that flanks the loch in part. Swallows and sand martins flicker over the surface. The water is clear; the blue of it ringed with wind ripples, and though we have passed no one to get here, a disposable barbecue lingers in a blackened square of soil. An empty packet of tattie scones lies in a burned-out drum next to a discarded Strongbow can.

It is eerie. The water is clear – oligotrophic – as the water generally is around here. But there is no visible life. No trout rising to ring the water's surface. No insects struggling in the surface tension.

The Loch Dee project was an attempt to correct the acidification. In the 1980s, 175 tonnes of powdered limestone was added to the burns that feed the loch in an attempt to raise the pH level, neutralising the acid for the sake of the fish and other life. It was a success: the pH level increased. But the problem lies in timing. Once lost, some things remain lost. What Barry Lopez called the syntax has gone. The limed loch lies in a sort of limbo.

A quirk of being a freshwater fish is that each loch's population is isolated from the others. It means that each population of brown trout is slightly different. The trout of Loch Grennoch (a little way southeast of Loch Dee) are genetically less susceptible to the effects of an acidifying loch than others and so it has been possible to use these local, acid-resistant fish to restore the lochs that lost their trout. The sting in the tail is that no one knows what was in Loch Dee that could have had an impact somewhere else in the future – its life forms might have had their own genetic specialisms, their own specific roles, their own future-proofed attributes. By needing to restore the water, you lose what was originally special about it.

We don't talk about acid rain any more. It is a problem left to high-school science lessons. An internet search will lead you to countless articles about how the problem was solved, how it is evocative of 1980s eco-fear like the ozone hole or nuclear disaster. The Scottish Government tells me that between 1990 and 2014, sulphur dioxide emissions were cut by 90 per cent and nitrogen oxides by 69 per cent.[27] Never mind that they are still emitted, it still rains and that what is up in the air still comes down. In 2021, the UK government's own figures showed 44 per cent of acid-sensitive habitats were still receiving inputs of acid over the level of harm.[28]

In Michael Wigan's book *The Salmon* – the one fish that perhaps trumps the brown trout for cultural importance to Scotland – he records a newly forested Galloway where 'Acidity levels were so high that when it rained hard the water in so-called "acidic flushes" actually suffocated adult fish with deadly aluminium. Young fish writhed in lethal water acids. Salmon eggs shrivelled in sour pools.'[29] It is an ecological quirk that in the Pacific northwest of the USA and Canada, salmon die in large enough numbers to supply nutrients to the trees, sustaining the life of the great forests, while in Scotland the great afforesting of Galloway killed the salmon.

Derek Ratcliffe, who kept returning to this corner of Galloway throughout his life, witnessing its transformation from a hard wildland of mountain birds and hill crofters, was moved to write, 'I believe afforestation in Galloway . . . has over-stepped the boundaries of acceptability by a fairly wide margin.'[30] As a scientist, he doesn't convey much emotion in his writing: the word 'fairly' is concealing a lot of feeling there.

Trees talks me through the plantations that ring Loch Dee. The one behind us is more recent and has been moved back from the water's edge in response to changing forestry practices, leaving the riparian buffer zone that we are sitting in, among willow and rowan, bracken and willowherb. On the far side, a block of emerald Norwegian firs and blue-tinged Sitka spruces is parted only by the flow of an unplantable burn. Once, these plantations would have run right up to the loch. Their gradual withdrawal is current government policy: plantations currently have to end 10 metres away from 2-metre-wide watercourses and 20 metres away from larger watercourses and lochs.[31]

I tell Trees about Derek Ratcliffe. He doesn't agree of course:

> In my mind, provided you can minimise the impact on ecologically sensitive areas, commercial forestry plantations provide a local renewable source of materials and jobs for thousands in the UK. The benefits far outweigh the costs for the most part. So whilst they may not be how I'd design a woodland, if I see larger rectangular blocks of woodland, I still see a good crop which provide so much more than the otherwise degraded pastoral land surrounding it.

Other people would disagree with that, as I do, but I let it hang. The day is beautiful, the scenery stunning even if the loch seems lifeless to our eyes today* and this might be the last spell of brilliant sunshine we get this year. Neither of us will change our minds: we are too divided by a common love of nature and different visions of the land.

---

* Brown trout have been restocked into it and anglers catch pike here.

He adds, 'I'm just surprised how many people see a conifer plantation and think it's natural. When people come to Scotland they think that's what it should look like.'

I agree on that point at least, with this loch whose water and sediments bear no resemblance to what it once was. The more we do to the earth and the air, the water will respond by degrees, turning away from what was natural.

An hour and a half's walking later we are back in the forest-flanked glen, beside Loch Trool, the rain bucketing down again, as if the day really wishes to make its point. We hide under pines while Trees puts his waterproof cover back over an already sodden rucksack.

The air above has been cleaned by the rain. What had gone up has now come down. It will find its way into the loch, directly or ushered on a carpet of pine needles into the nearest burn that enters the loch. The loch will go on, looking natural, lying to us. But the dark water hides depths, the fingerprint of our mistakes, histories that we might rather forget. Because no water remains still and unchanging.

I lean back. The canoe sways gently below me. Through a window of cloud, two birds drift, their wings bending towards the tip, each finger-feather held out like a ballet dancer's hand. The air caught, the gentle breeze sends them slowly pirouetting through the blue sky. Ospreys. Their pale breasts shine white as clouds, the shadowed wings black. Their bent wings make them look as though they're constantly soliciting a hug, as if ready for the loch's embrace.

They nest somewhere, I don't know where, in the wallpaper of pines that smother the ground between the Galloway lochs. Irregular enough that they remain the highlight of a day. This pair carry on their slow dance through the sky, high, able to survey all the water that glimmers through the trees like shards of black glass. A fish is the target. Below the layer of glass, the shadowy presence that they can see through glare and moving waves; appearing as elongated ovals, piscine pills.

This is the Galloway loch cycle. From zooplankton to drifting mayflies trapped on the surface of the loch; fish feed at every level in the drifting currents of the loch before – ach – one false drift through the water column and it is snatched, the exquisite pain of a talon in the back, crushing, the wrong oxygen passing through the gills, drowning in dry air.

Today the ospreys draw a blank and drift elsewhere. Years past I saw it; a different Galloway loch, ringed with pines of course and an osprey, folding its wings back and falling fast, feet first, into the water. A momentary pause, the bird held by the arms of the loch; then a flurry of wings as it lifted up, a flurry of droplets down and in its talons, a fish flying into the sky.

5

Green as a Dream

Waiting. Our droplet was swept up from a loch and drifted in a different direction, to the south and east. It fell months ago but the journey did not stop. Instead, it patiently travelled, filtering down through the porous chalk rock, stripping away the sky's impurities from the perfect water. Squeezed for miles along fissures, it moves with a slow yo-yoing, up and down and sideways, through the earth, until it pushes through, risen somewhere new; born again, in southern England, into the light of a new stream.

In the chalk stream of the mind, it is always a languid, late-spring afternoon, the sun always shining on a water that flows, cool and clear. They are rivers but note the slippage in terminology: diminished to stream, a suggestion of their smaller size, their lower energy, their dreamlike state. Note also the chalk. For it is the rock beneath that the water is stored in, purified by and pushed through, that makes the chalk streams of England and northern France unique. It is this, their particular combination of water and land, that makes them special, that makes them seem like *The Wind in the Willows* come to life.

Chalk streams don't carve their route. These are not places of a vulgar energy but of gentleness; the water seeking out the lowest point in the landscape and then ushering itself down, gliding between lush banks, lined with the feathery-leaved willows that flicker between green and grey and green with the drifting of the breeze. The drooping branches sway above the flowering white froth that waxes and wanes with the passing months; first cow parsley, then hogweed, then lesser water parsnip; each flower covered with insects, seeking out the pure sugar at their heart.

Below, in the flow, emerald weeds drift sinuously in the current. Water crowfoot blooms clear of the surface with a white buttercup-type flower, like a lily in miniature. Below the weeds, white-clawed crayfish scuttle, like small, squat lobsters over the gravel bottom, from hiding place to hiding place, grabbing at anything edible. It might be fish eggs or decaying plant matter. It might be invertebrate larvae as they begin to feel a stirring within; responding to time and temperature and moving towards the banks, where they crawl up the vegetation, clinging to safety above the water's surface. Then they split their own skeletal skins open. Squeezing out their new soft bodies, as wings unfurl and harden rapidly in the sunlight. As sulphur-yellow mayflies and dusky brown caddisflies emerge, they turn the water into a rippling carnival, the surface ringed with the movements of trout, sucking down the flies that don't make it out of the stream's surface. White-legged damselflies drift close to the banks, seeking out the small insects that escaped the water's surface. Scarce chaser dragonflies shoot high over the flow, never seeming to land, restlessly seeking out the bigger insects to eat in the short summer blaze of their life on the wing. As they hunt overhead, salmon seeking sex have followed their noses back from the sea to the gravel bed they were born in; to begin their cycle again, as a milt-fertilised egg wedged in the gravel, washed by the river's flow. One cycle in the flow of a larger cycle: the water's constant movement from deep underground to the sea and the air above and the flowers beside.

They compel you to think of them this way: as rivers of life; pristine waterlands; the earth emanating its cleaned and polished plasma. Because these are rivers that just have something about them – some magic – that others do not. As Rupert Brooke wrote about one: 'The stream mysterious glides beneath, / Green as a dream and deep as death.'[1] But to describe the chalk stream from the ideal is like describing southern England as an Eden. Most of our chalk streams have fallen. Still green but no longer deep as death; now shallow and gasping for breath.

'We'll start with something depressing,' says my cousin Lizzie, cheerfully, as we leave the car behind and walk down a tight country lane to a fold in the valley floor. Lizzie works with degraded chalk streams, an interest ignited by studying rivers for the same A level in Geography that bored me senseless fifteen years ago. In the arid mid-Suffolk of my childhood I found it all hypothetical, whereas she had her relevant examples on hand in Hertfordshire.

The scenery we pass through is not exactly wrong: dark green oaks and paler hawthorn hedges line the faded fields of green and brown that creep up the sides of a shallow valley; pastoral southern England in deep summer. And yet it is not wholly right either. Chalk is here but it is hidden: there are no white scars in the landscape, no pale stones kicked up by the field edges. It is deeper below us in the earth and the water.

We walk to where a fallen willow has shed its branches over a ford in the road. The crossing is dry. A bridge brackets the road, elevated by a foot or two over a ditch. 'It's completely dry,' says Lizzie at first glance, a tinge of shock creeping into her voice. 'I was here a month ago and it was running,' she points from halfway up one bank of the ditch to the other, 'a good few metres wide.'

Not completely dry: at second glance there is an inch of water puddling, barely flowing between the thick green growth of *Glyceria*,* the various thistles and the lilac flowers of water mint. In the middle of the channel, dung flies, like buzzing flecks of amber, gather on the paradise of a cowpat.

I'm grateful to have her eye and experience with me because I would not have recognised this as the River Beane, a chalk stream, just as I never recognised the River Lark, running in a concrete ditch through the car park of my local Tesco, as a chalk stream when I was at school. I thought it was a drainage ditch. I would have said the same about the Beane, nothing special, not an example of England's great contribution to the world's rivers: 85 per cent of the world's chalk streams are found here.

'It is completely straight,' Lizzie says, explaining the issue with this scene, from her point of view. 'Do you see there are lots of plants in the channel? These are marginal plants. They should be growing on the banks but the reason they are growing in the channel is because of too much sediment being deposited by low flows and the modified morphology.'

In the horse paddock beside the Beane we walk to a depression in the field, a wriggling snake of lower grass with clearly defined edges: the relict channel. Nothing else remains here of the Beane as it was before its new course was dug, many years ago, along the straight edges of two fields, funnelling the water away, allowing all the possible land to be wrung dry for grazing. There are no river flowers, no riverflies, no water, no silt or stones left to suggest that this was the Beane; just the gentle ghost of its shallow bed, grass where the water once was.

None of this is as it should be.

*

* *Glyceria*: sweetgrass, its green leaves that grow up are intermediate between an iris and a reed.

To know what should be special here we need to begin, once more, at the real beginning.

A hundred million years ago. Pangea has split but the pieces of the continental jigsaw have not quite fallen in their final places. Though even if it had, we would not recognise our Earth yet. After all, *Tyrannosaurus rex* walks what will become western North America, *Iguanadon* stalks southern England. The climate was hot – both poles were ice-free – and shallow seas crossed continents. One such sea covered southern and eastern England and northern France. Within the sea coccolithophores (phytoplanktons with calcium-rich exoskeletons) proliferated, turning light into oxygen and life.* When these organisms died, their skeletons sank to the bottom of the sea; the bed of dead coccolithophores accumulated by a centimetre of depth every thousand years. They lay there for millions of years, gradually compressed into rock under their own weight and the pressure of the sea. They remained undisturbed until the end of the period, when an asteroid arced through our atmosphere into the Yucatan peninsula of Mexico and changed Earth again, completely, ending the Cretaceous. From the Latin, *Creta*, meaning chalk. The age of chalk: the rock that was once alive under the same light as the dinosaurs.

The earth-bound dinosaurs died out with the asteroid, leaving only those that would become the birds. The ancestors of the dragonflies, damselflies and mayflies survived too. The asteroid did not work alone but in tandem with huge, planet-cooling, volcanic events. A cooler Earth is a drier Earth. The sea level retreated, draining the shallow warm seas, leaving the newly formed chalk beds across Europe. When the continents collided into their final places, the shockwaves lifted the chalk of England and France up, from

* The diatoms of Chapter 4 are a type of phytoplankton, only with a glassy rather than calcium-rich exoskeleton.

seabed to the surface; chalk elsewhere in the world lies covered by the oceans still.

Aeons after the Southern Uplands of Scotland had begun to gush water, this corner of England and France followed.

When rain falls here, it does not flow into the river immediately, as it would elsewhere, colouring the water with the land's sediment and detritus. For chalk is porous and can hold up to half its own weight in water, and so the water travels down rather than across, into the aquifer rather than the river. It can flow underground and re-emerge in a chalk stream many months and many miles away from where it fell to Earth, its journey through the alkaline rock neutralising any acids, scrubbing away the impurities it might have picked up in the atmosphere.

When this water gets released slowly back up through the fissures in the chalk, it emerges as springs of a consistent flow and at a constant temperature of 10 to 11°C – the earth's temperature – meaning it is cool in summer, warm in winter. The water is as clear as a cut diamond and almost as valuable, as different from our other rivers as flowing water can be. Because of this, the Environment Agency and Natural England, otherwise sober organisations, have referred to them as 'England's rainforest'.*

That is the ideal, how chalk streams should work. But then there are other chalk streams, tangled with other geologies, other rivers, hidden from both the ground and the sky in culverts below towns. Like the River Beane, we have moved them; ripped them from their groundwater and rerouted them to where we want. We know what a chalk stream is by when, where and how they were formed but we can't necessarily agree on whether these trickier ones still count, if they should fall from their hallowed status. Because of this, the

---

* This cliché ignores the fact that England already has temperate rainforest, in Cornwall, Devon and Cumbria.

exact figures are disputed but the Catchment Based Approach* has a list of 283 chalk streams flowing in a fine lace across southern and eastern England. Always shallow and not particularly wide or long, chalk streams co-exist with normal rivers – often flowing into them – and they are imperilled in the same way. But they also face different threats. It is their misfortune that the chalky south and east of England are the driest and mostly densely populated parts of the country.

The local water company for Hertfordshire, Affinity Water, takes between 60 and 66 per cent of the water it supplies from the chalk aquifer.† What flows through a tap and flushes a toilet in this county is more than likely water that would otherwise have fed a chalk stream: to extend the rainforest cliché, that's like chopping down the Amazon to produce two-thirds of the paper that you are reading this on. Water companies do this because in 1945 the government passed a water act that limited the amount of water that could be taken directly from rivers. It left a shortcut. Aquifers were unprotected by the act, and so companies began to take it from there instead. This loophole has never been closed (merely licensed, to ensure 'responsible' use.). On the nearby River Ver, 45 million litres a day were abstracted from the aquifer in the 1980s. In a dry spell this was all the water that would reach the aquifer from a year of rainfall.[2]

Chalk streams are supposed to be connected to the groundwater: they flow where the earth is saturated with water. Removing groundwater puts the brakes on a chalk stream. Rivers flow where they want to flow, in the perfect place for them – if not us – and any changes to that flow is deleterious to them. The re-routing of

---

* A loose group of organisations, ranging from DEFRA and water companies to the RSPB and the Angling Trust, all of whom have an interest in the waterscape.

† Various figures of 60 per cent, 65 per cent and 'two-thirds' are available on its website, affinitywater.co.uk.

troublesome chalk streams out of their natural beds into artificial channels severs them from the groundwater too and leaves them 'perched', above the lowest point in the landscape. The groundwater will still bubble up in the original bed, even if it is lost to the chalk stream. Water, self-willed, will remain troublesome even if we try to manage it by brute force.

Abstraction has two meanings. In the technical sense it means to take or remove something. In the more commonly used sense, it is about ideas rather than a true representation (as in the way that abstract art doesn't strictly represent something). The River Lark, in its concrete straitjacket running through a car park, is a chalk stream that has been made abstract, taken out of its chalk and clear water reality. It is a tragedy. Water in southeast England – distant rainfall, purified by a chalk aquifer and given up to the surface through a spring – has become abstract to us too. It has instead become a convenience: a thin, fast flow from a tap, a freshly boiled kettle, a flushed toilet. (It is also a curse: the chalk minerals make 'hard water', resulting in limescale and brittle, dry hair.) The ease of this connection obscures what water is: it is hard to relate it to the beginning and end of the water cycle. Before privatisation water was 'treated not as a commodity but as a prerequisite for public health'.[3] It is harder still, now that this substance, this stuff of life, is treated as a commodity, managed for profit and not for the benefit public or places. The meaning of water has been abstracted, while still we bleed the ground dry.

This is not the only way we are harming our chalk streams. As well as removing water, we are introducing chemicals. A result of this is that English rivers, not just the chalk streams, are in trouble. If it is shocking that we are somehow in a situation where only 14 per cent of English rivers are in good ecological condition,* then

* The same percentage as it is for England's lakes.

it is horrifying that none – zero rivers[4] – reached the threshold for being in good chemical condition.* The water that the chalk aquifer provides is supposed to be special for being pure. The nutrient paradox we saw in the lochs rears its head again. In water that has little chemical content, any added phosphorous or nitrogen immediately has an effect, turning the water sickly green, altering the chemistry, removing the oxygen. The low-energy nature of chalk streams works against them here: without the riffles required to oxygenate the water, pollutants can literally choke them; when the weather is hot, the pollutants remain, increasing in strength as the diluting river water evaporates, the groundwater dries out and the river level falls.

Typically the most tangible source of these chemicals comes from the sewage network: leaky sewers, storm overflows and treatment works. An investigation by Unearthed, the journalism wing of Greenpeace, discovered that in 2022 'over 38,000 hours of sewage were released into or within 50 metres of chalk rivers in England'.[5] We are defecating into our own drinking water; wiping ourselves with the chalk stream rainforest.

There are other chemicals and other sources. F. W. Berk & Co. was a chemical company. At a site in Sandridge, just northeast of St Albans, it manufactured bromine-based chemical products for twenty-five years until 1980, and at some point bromide and bromate seeped into the groundwater. After production stopped, the buildings were demolished. A few years later, after bromine was discovered in the water, the topsoil was excavated and a housing development was built on top. It didn't solve the problem. By 2000 bromate was discovered in the groundwater below a pumping station

* Down from 97 per cent passing in 2016. It's not that the rivers are getting that much worse, it is that we're getting better at looking and detecting things that previously went unfound – PFAS in particular.

in Hatfield, 5 kilometres away. Subsequently this plume of bromate was discovered to have affected an area 20 kilometres wide, percolating through the permeable chalk, polluting the aquifer. After a quarter of a century, efforts are still taking place to fix it, although this is nearly impossible in the diffuse water of chalk. The pumping stations near the plume no longer supply drinking water. The water they pump is predicted to stay above the water-quality threshold for bromate for the next 200 years.[6]

This isn't an isolated occurrence. Permeable chalk is more vulnerable to this sort of pollution than elsewhere: the knock-on effect of that groundwater being taken out of the drinking-water supply puts more pressure on the clean aquifers elsewhere in the rapidly drying southeast of England.

The Beane is not wholly depressing. There are parts that still resemble what it used to be like, what it could be again. Lizzie has taken me a few kilometres downstream to Waterford, a village on the edge of Hertford, where mansions and old cottages sit beside willow-fringed green spaces, and new-build terraced bungalows are hidden down dead-end lanes. A nature reserve brackets the east bank of the Beane here; a brown swathe of long grasses, thick with purple flowering thistles that conceal shallow pools. The path meanders with the river. Willows droop to the water's edge, which flows clear and with frequent patches of gravel on the bed. Stands of great willowherb and marsh woundwort run lines of pink parallel to the river. Banded demoiselle damselflies flicker – a dark teal shimmer – and a kingfisher whistles from upriver. Lizzie takes her eye off the water for a second, the exact second that it chooses to shoot low over the river, a vanishing blur of blue. Bad luck.

A river is always a work in progress. We're here because Waterford Marsh is a particular work in progress, showing how a Hertfordshire chalk stream could be rescued. At the start of the year the Environment Agency collaborated with Hertfordshire County Council on restoration works. The old riverbank, a crumbling concrete lip that hemmed the Beane in, was removed. The Beane's banks returned to the earth. While they were there they also installed 'brash berms', a wide line of woody material by the bank, and submerged lumps of wood to deflect the flow. The two works are complementary. Concrete is uniform, which a river does not want to be. To be a living river requires interruption to the flow: calmer water alongside fast, back-currents and eddies and slack water; a river is, crucially, not a road. Installing berms allows marginal plants to return, to gain a foothold by the water's edge, where they help keep the earth in place. It also deters dogs – and this is a serious problem that does not sound serious – from entering the water, where they can have a profound impact eroding the banks, trampling vegetation and washing off flea treatments that often contain strong chemicals such as fipronil, a pesticide banned for agricultural uses due to its impact killing insects in the environment. Submerging trunks of trees in the water has the same effect as the berms but works better for aquatic insects and young fish. It gives the insects a habitat and young fish shelter from the main flow – and provides them with the insects as a food source. 'Fish grow on trees' is a phrase regularly used in hydro-ecology.

These works are seven months old. It's no time at all by which to judge their effectiveness. But as we walk along, fish fry dart and wiggle in the clear bankside shallows. Azure damselflies part the gloom of the shadows like a needle sewing a sky-blue thread. A moorhen delicately picks its way over the branches of a berm. A few water crowfoot flowers are held above the water surface, like a beacon of health. It almost looks right. Not perfect, but in the clear water a gravel bottom can periodically be seen: perfect for the trout and grayling that have been lost but might one day return.

Problems still exist. Further downriver we pass a vast stand of rhododendron, a Himalayan shrub imported for garden planting. When it escapes it grows densely in the wild, starving the undergrowth of light until the plants below die. The bare soil is then flushed straight into the river during heavy rains. Also present is Himalayan balsam, a plant with a purple and white flower that grows to head height, which has a similar effect. It spreads by shooting its seeds along the riverbank from a 2-metre height, flourishing to the detriment of everything else. Fewer flowers in the undergrowth leads to fewer insects which leads to a less diverse, less healthy stream. At the end of this stretch, the water suddenly becomes cloudy and constrained; an old weir and sluice interrupt the stream, preventing fish movement, while the banks either side are suddenly thick with dense greenery, choking out the light from the channel.

And yet. We turn around. Walk back into the damselflies, past the river plants, to where the water is clear and the evidence of the eye says corrections can happen. That this beautiful stretch of river, still with life, is worth holding on to and replicating elsewhere along the Beane.

England's premier examples of chalk streams – the Test and Itchen, further south in Hampshire – show up on maps as an unravelling braid of blue threads, bunched and branching off through the surrounding floodplains.

'Chalk streams are supposed to be connected to their floodplains,' explains Lizzie. 'They're supposed to have marshes, wet woodland and floodplain meadows around them. They need a lot of room.'

But historically we are not good at giving rivers room.

The posh houses at Waterford lie at risk within the floodplain (while, perhaps unusually, the terraced bungalows don't): the lure of the pretty view of river and willow trumping common sense. I have seen pictures of Waterford Marsh in winter after exceptional rainfall, when the common land next to the river has been entirely flooded. It is a world away from a summer's day like today – but the river is here all year around, even if we're not. The edges of a watercourse tend to be ephemeral anyway, expanding and contracting in response to weather conditions; dry summers and wet winters. But just because it is ephemeral doesn't mean it isn't important. It probably makes it even more important.

This is recognised elsewhere in the world. A different model we could follow.

'In France when they think about a river,' Lizzie tells me, 'they're not just thinking about the channel itself, they're thinking about the rate at which it can expand in a flood event, so that's inclusive of the whole floodplain. That is the river and that's what's protected as the river.'

More than half of the departments of France – roughly equivalent to an English county – are named after rivers; the name of a French river can apply to the watercourse, the catchment and its inhabitants. They respect their rivers and are sensible with water. The *bassin-versant*, the watershed, is a powerful concept, organising how the French administer their water resources (as we used to). Now Affinity Water, owned by a German financial-services company, handles the water supply across a broad swathe of Hertfordshire, Buckinghamshire, west London and Surrey, as well as parts of northeast Essex and southeast Kent – a lot of unconnected watersheds.

'Here,' Lizzie laments, 'we have a habit of treating the channel as separate from the adjacent land. Actually, rivers are supposed to expand and contract with weather events. And we're trying to contain and stop that but they're not meant to stay in their banks. If you try to keep them in their banks, they're not going to be very happy.'

I love her phrasing – a glimpse of the emotional lives of rivers; the truth that all living things need connection.

The worst thing we ever did to water was try to control it. Lizzie has somewhere else to show me that demonstrates what can happen.

We have come to the Woodhall Estate, a little way back up the Beane. The day is closing in, the humidity stifling; the fields wanting the rain that the sky is holding back in gloomy, overcast clouds. We trudge between an overgrown verge and a fast road, field bindweed returning the path in between to green; the purple-blotched stems of hemlock in the verge bleaching, turning straw-coloured. Through a gap in the hedge we can see the estate's fields, the grass grown long and fading to an arid-looking sandy brown. A line of willows runs through the field beyond. A suggestion of the water that the map says is there.

Hedge gives way to brick wall. By the main entrance it becomes elaborate: stone balustrades opening the landscape up, leading to an ornate brick lodge and an opening in the wall. A narrow ribbon of tarmac runs down into green parkland between two thin iron fences. On the far side, sheep graze the grass that runs up the slope to the manor house. In the middle, a broad sweep of water: the estate lake, an essential part of any eighteenth-century parkland landscape gardening. It is the River Beane dammed and diverted, a chalk stream turned into a lake along the valley floor. In 2016, it was breached by an intense storm; excess water caused a bank to subside, draining the lake into the Beane, leaving two centuries of accumulated silt – 25,000 cubic metres – in its place.

The broken bank was repaired, the silt dredged from the bottom of the lake. But it was an opportunity in disguise: the estate worked with Affinity Water on a restoration programme that aimed to prevent this happening again. A bypass channel was dug, in essence offering water the choice: to flow into the lake or to flow around the lake. Most chalk streams flow through multiple connected channels anyway so this is closer to the natural process than it may sound. Or look.

We have walked to the bottom of the road. The bypass channel had been tucked into the folds of the landscape but at the southern, downstream end, where it joins together with the lake outflow to recreate the River Beane, it is unignorable. Two banks of clean, pale brick step down, a glitchy pixel version of the gentle curve of the land behind. It is neat – as if landscape gardening was solving the ecological problem caused by landscape gardening – and it works: water flows down it, joining the lake's outflow, the water ruffled and oxygenated by the merging of the two streams. Although it is not totally clear, it is flowing water. And it is clear enough that a pale movement counter to the flow is visible, flashing through the water.

Moving water is hard to look at through binoculars. Using them to see something you've already spotted with the naked eye requires something solid, a marker to indicate the right direction. With the ever-shifting shadows of flowing water, you don't have that. I train my lenses by the ripples between flows, halfway up the shadow from the trees opposite, rotate the focus to try to see through the surface, searching. A flash again: that outline, familiar from pictures, streamlined yet thickset; the flanks delicately speckled; the fins surprisingly small. A brown trout moving gently through the light, back across a window of clearer water, before vanishing. Visual ephemera. Like a ghost, glimpsed in the shadows of the stream. A ghost of chalk streams past. They are the essence of these rivers, feasting on the supposed abundant life. That there's at least one here is a good thing. Clearly not all is lost in the Beane.

Spurred on by the state of our chalk streams, and an awareness of how special they are, the Catchment Based Approach (CaBa) has come together to rethink the way that water is managed in England. It involves an appreciation of the French way of seeing a river beyond the banks of the flowing channel and towards the wider area: where water will go and from where water will move into the river. CaBa's approach uses a 'trinity of ecological health', a triangle of water quantity, water quality and habitat quality, each corner of the triangle inextricably linked. One of its proposals was to ask each water company with a chalk stream in its supply area to return one to a good status with ten years. Affinity Water chose to commit to restoring the River Beane.

The main goal doesn't sound ambitious: making sure water flows along sections that aren't winterbourne (those that flow only after winter rain) should be the bare minimum goal for any river, let alone those that are 'England's rainforest'. But it is something that the Beane is barely achieving in its upper reaches. The company's other aims are aligned with the trinity: manage the banks for chalk-stream species and make sure the water is clean. Again, you wonder how we ever got in a position where such aims are seen as a challenge and not the minimum required standard for a river: the rivers might be placid but we don't have to be.

This is slow progress. The preliminary assessment has just been finished at the time of writing. By the time you read this the working group partnerships will have been established. Between 2025 and 2035 the work will be undertaken. I guess it will show if they actually have an affinity for the substance they supply in its natural form. It can be done: Waterford Marsh shows one way, Woodhall shows another – a more creative solution to disasters.

More importantly, it *must* be done – the damage we have done won't undo itself.

'Chalk streams are hostage to their own placid selves,' says Lizzie. 'They can't overcome the challenges they face when it comes to modification. So if you were to put a weir in a high-energy river, it might eventually bypass it. A chalk stream won't be able to, because it is so low energy. So you need human intervention to solve those issues, which is a really big challenge in somewhere like Hertfordshire.'

'Where there's so many people?'

'There's such high population density especially in the south of the catchment, you've got no room to put them back to what they should be, so it is really difficult to undo the harm we've done to them.'

This has to work. The Jaws of Death is a graph that all water companies have. One line plots the increase in water demand. Another line plots the available water to supply those needs. In most graphs these lines cross each other in 2050. Already in water-stressed Suffolk, developments have been refused permission for their impact on water resources. The deficit between England's sustainable water supply and demand is forecast to be nearly 5 billion litres a day by 2050.[7]

If we can't keep the water we do have flowing, what sort of future lies in store for us?

I wanted to write about chalk streams as a refuge of rare purity, the idyll that I imagined at the start of this chapter; or the chalk River Lymm-inflected images that Tennyson uses in 'The Brook', where:

> I wind about, and in and out,
> with here a blossom sailing,
> And here and there a lusty trout,
> And here and there a grayling,
>
> And here and there a foamy flake
> Upon me, as I travel
> With many a silvery waterbreak
> Above the golden gravel,

And draw them all along, and flow
To join the brimming river,
For men may come and men may go,
But I go on for ever.[8]

Bill Condry took nature writers to task in 1968 for 'falsifying the picture'[9] and I agree that there's a moral responsibility not to pretend things are fine.*

Water might be impossible to destroy physically but that doesn't mean that a river can't be destroyed, its flow throttled, its life ebbing because of the things that have been done to it. Tennyson's brook is a metaphor for life in its intertwining of certainties and change: as a metaphor for life it still works. But the language seems quaintly old-fashioned for a river now, when it is more likely to brim with shit and dry up in summer; while water company shareholders come and go.

If the first blow was landed in 1940 with the aquifer loophole, a crucial series of further hits happened in the 1980s. Water – drinking, sewerage, conservation – was organised separately through local authorities and companies. This responsibility was then nationalised by Ted Heath's Conservative government in the 1970s and grouped into ten boards, the Regional Water Authorities, bringing together all the concerns that were hitherto separate. Each authority functioned with its own area's needs in mind, presciently realising that the southeast of England might have quite different requirements from the northwest or Wales. However, by 1979, with the election of Margaret Thatcher's Conservatives, things changed. The Regional Water Authorities were suffering. According to Professor Karen Bakker's academic account, 'Labour levels and investment were

* Here Lizzie would like me to make clear that what follows is purely my own opinions and not hers or her employer's.

reduced, tight financial controls were introduced, price increases were mandated (with bills rising above inflation), and increasing emphasis was placed on economic, as distinct from technical, performance indicators.'[10] After other public utilities including British Telecom and British Gas were privatised, the Regional Water Authorities were next. In 1989 the ten were sold off as private companies.

To give them a head start, the government wrote off their debts – a figure that, with an accountant's exactitude, equalled the £7.6 billion that the government earned from their sale. So we sold off our water for free. The aim was for the free market to demonstrate its superior efficiency: its ability to invest more, to move more nimbly than the government could. In December 2023 it was reported that water companies were £60 billion in debt. Thames Water, the largest of our water companies, handling the tap water and sewerage for 16 million people from Gloucestershire to Kent, accounts for £16 billion of that debt; at the time of writing it is not clear whether the company will still exist by the time you read this. Debts have to be repaid. Reportage by the *Guardian* has revealed that for three water companies – Thames, Southern and South East – over a quarter of their revenue (your water bills) goes towards paying this debt.[11] For Affinity Water it was just under a quarter.

The problem is that water – in nature and geography – is not a good business model. One cannot simply dislike Thames Water and be plumbed into another company's, another catchment's water supply instead. The infrastructure of water requires maintenance and monitoring. It needs investment. Professor David Hall from the University of Greenwich's Public Services International Research Unit is a strident advocate for re-nationalisation. In a readable, punchy paper in which he rebuts the industry's claims, he merely notes that Scottish Water, a publicly owned company, has invested more money per household than the privatised water companies, charges lower bills, does not pay dividends to shareholders and is reducing its debt.[12] Even Dŵr Cymru, Welsh Water, one of the Regional Water Authorities privatised in 1989, was bought as a debt-ridden company for £1 in 2000. It is now a not-for-profit organisation.

The problem with profit is that it disappears in dividends for shareholders. In Hall's analysis an average of £1.6 billion becomes dividends each year for water company's shareholders. Some will pay more, some less, but each payment is money not reinvested into our water infrastructure and environment. The balance sheet then allows for concepts such as the Economic Level of Leakage, by which a company calculates the expense of fixing leaks against the expense of losing water to leaks. Companies then fix the leaks that save them money. Their interest is in money rather than water, so even in droughts or in regions of water scarcity, such as the southeast of England, it is easier for them to tell people to use less water rather than take better care of it themselves.

Privatisation has been a disaster in slow motion, the money in the system skimmed off as profit, and now English water bills go towards paying corporate debt instead of fixing leaks. Our public health has become a money-spinner. Our environmental health too, since with water the two are inextricably linked. Hall, on an episode of Radio 4's *Costing the Earth*, described our current state of sewage pollution as the 'economic level of sewage'. Perhaps this is how we should be seeing our chalk streams as well. Over-abstracted, pungent with pollution, barely flowing: the economic level of a chalk stream, not necessarily going on forever.

It's August now. Another muggy day, desperate for the clarity of a cool breeze. The town of Stockbridge sweats money. The car parks heave with 4×4s, dull-green gilets and posh accents through the open door of the deli. A carved wooden trout hangs above the door of the town hall. A golden trout glistens from its weathervane. Pick-up trucks clad with the logos of trout-fishing companies set off, fully laden, for an afternoon's guided angling down private lands. Places that aren't meant for the like of us.

This is the River Test.

I had to come after Lizzie apologised for the Beane. If that was the common reality, I wanted to see the outlier, though this is harder than it first seems. The River Test is aggressively private: English landownership laws allow this jewel in England's river crown to be kept mostly away from the rest of us, in case we accidentally fall in love with it without paying for the privilege. There is money to be made in keeping it for those who can afford to pay for exclusive fishing, rather than letting walkers look at the wildlife along its banks and the water within for free. We are left with little scraps.

The town is long and thin and cuts across the valley that the Test runs through. Channels of it – those unspooling silver sunlit threads of chalk-stream water – slip through the town, surfacing in gardens, between buildings, leaving a culvert under the A30. We park up by the culvert. In shallow water moulting mallards wait for crusts of sourdough bread sandwiches. Beneath them a brown trout lurks, a speckle-backed bar of bronze in water like liquid glass. It is still in the flow – poised, waiting, ready.

'Decent size. That's about 3 pounds, I'd say,' estimates my brother, Chris, a coarse angler* with an appreciation for fish that extends beyond the species he targets. He flicks a bit of a sausage-roll pastry to it. In a split second the still trout reacts, snatching the pastry from the water's surface, quicker than the ducks can move.

---

* Coarse fishing means fishing for species that aren't a food source. It is cultural definition. Other European nations eat species of fish that we don't often touch in Britain.

'There's another, a rainbow,' says his wife, Hannah, as she leans over the railing, one hand left on the pram rocking their two toddlers. A thinner, less muscular-looking trout appears at the end of Hannah's point; delicately spotted with a pink line running down its flank. The rainbow trout is a species from the Pacific, introduced to British waters for fishing sport. They don't impact the brown trout currently because they never established significant populations: nowadays only triploids* are stocked, which are sterile; their presence in our rivers a temporary, deliberate thing, fuelled by the anglers' desire for a thrill.

Further down the road, the main channel of the Test runs under the road. No, 'runs' is the wrong word, a verb for other rivers. The Test glides, effortlessly, unflustered, the leaves and stems of water crowfoot swaying to the beat of a flow that has pace but doesn't break the perfectly smooth surface of the river. Half the channel is dense with its growth; the other half shines with the bony chalk beneath, a backdrop against which trout stand out, obviously, evidently with some confidence in their own safety.

When Roger Deakin passed through here a quarter of a century ago on his *Waterlog* journey, he reported smaller-looking trout, in a twee, eccentric village. Some things change. But other things, as he discovered the day after on the River Itchen, 15 kilometres south-east of here, don't. After swimming in the clear chalk water of the Itchen he was assailed by a porter and a river keeper threatening to call the police. Deakin and I are different personalities with different priorities. He wrote: 'The truth was, I had enjoyed my row with the water bailiffs very much. I already felt invigorated after a really first-class swim, and now I felt even better after a terrific set-to.'[13]

* Triploids are fish that as eggs are treated with either heat or pressure to ensure that they retain a third chromosome instead of expelling it.

Whereas I am cursed by being conflict averse. But it leads Deakin to a point with which I agree:

> That so many of our rivers should be inaccessible to all but a tiny minority who can afford to pay for fishing 'rights' is surely unjust? I say 'rights' to point up the paradox, that something that *was* once a natural right has been expropriated and turned into a commodity. Fishing rights are only valuable because individuals have eliminated a public benefit – access to their rivers – to create an artificial private gain. The right to walk freely along riverbanks or to bathe in rivers should no more be bought and sold than the right to walk up mountains or to swim in the sea from our beaches.[14]

I couldn't put it better.

Chris, Hannah and I wander south down an overgrown path on the edge of Stockbridge, lifting the pram over fallen thistles and past the grab of stinging nettles, to a gate with a sign that reads: 'CAUTION! Entering the water is not advised', before explaining that the area is affected by 'intermittent sewerage discharges'.

This might not be the River Test after all.

To be technically correct it is the Marshcourt River that the map declares as running alongside Stockbridge Common, just below the town. But this is a chalk stream too. The maps are unclear where it begins and where it becomes separate from the many channels of the Test before it joins back up with the main river 2.5 kilometres south of here.* Water in the porous landscape of chalk defies our

---

* The National Trust, which manages the common, refers to it only as the Test. Its website says, 'The marsh offers access to the river's edge, in a valley which otherwise has little or no river frontage.'

strict definitions. The common is public access and initially there is no water to be seen, only felt in the soft peaty soil that quakes under my niece's footsteps, while my nephew sleeps. All there seems to be are chrome lines of ragwort, cream meadowsweet and long-grown grasses turning brown; the purples of knapweed (like a thistle exploding), loosestrife and great willowherb; the fluffy heads of hemp agrimony and stands of willow and—

'DON'T STEP ON THE COWPAT!' Chris shouts after his daughter. She steps on the cowpat.

—then a break in the undergrowth. After a few minutes' walking, water reappears. A small pool curving around the base of a willow. Green grass and dark soil are abruptly severed by a white line of loose chalk that crumbles down to the water's edge. The level is low. *Glyceria* erupts from it like a miniature reedbed; forget-me-not creeps around their bases like an afterthought. A heron gets bored of fishing and creaks into flight. Chris catches up with me.

'Oh, there's a trout.'

He spots it through his polarising glasses, skulking under the fringe of willows, a small, speckled, slip of a smolt-sized trout. I see a rough outline. I pull my own polarising glasses out of my pocket and feel the cheap plastic frames break in my fingers. It makes no difference. The trout zigzags off through the water, a startling burst of pace before I can find it again. A wilder trout than by the road, nervous, not waiting for food. The willows screen the water off from the common's edge. The path turns to flank the main course of the river but the bankside buffer of greenery – the tall profusion of willowherb and dense *Glyceria* – work to hide it, its presence only shown by the flora of the river edge that has grown high and shaggy, the vagueness of high summer in the moment where everything has lost definition at the apogee of its growth.

We get glimpses intermittently of the river, sometimes side channels separated by willows, sometimes where watercress has grown in green swathes, bedding down in the channel. Then the bank opens out. And Rupert Brooke returns to me. *Green as a dream and deep as death.* The willows on the far bank are like a wall, an impenetrable screen of greenery. The crowfoot swirls at the surface of the water that tugs at it, ceaselessly but without that vulgar effort; the water still clear, simultaneously present yet barely there. A plasma across which azure damselflies skitter as if trying to lure the trout into rising; banded demoiselles, their four wings marked with a smudge that flickers over the river at a slower place, a more graceful flight, their bodies dangling behind them like a fairy in blue-green. They land in the bankside vegetation. Above them gatekeeper butterflies patrol the hemp agrimony for interlopers in a feisty blur of orange. A sedge warbler chatters away somewhere in the impenetrable green of the place.

I feel compelled to pick my way through the barbed-wire fence. My feet follow an animal track to a lip of grass on top of the bank. I kneel by the riverside and dip my fingers in, feel the cool silk of the water flowing past my knuckles. This is the river you dream about. I feel it pulling at my heart while my entranced eyes check the shadows of the water, primed for a wild trout, speckled and golden and muscular, as if to say this – this arterial flow of purity – is proof that we can have the ideal, the perfect English chalk stream, intermittent pollution be damned.

Chalk streams have won their arguments. Everybody wants to protect them. There is the will but even more crucially some money for some of them. Recently the Norfolk chalk stream, the River Glaven, received £130,000 for its improvement; the Chilterns chalk streams got £350,000. Only time will tell how far that stretches – whether the fallen chalk streams, like the Lark in its concrete cage in a Tesco car park, will see any of that money, any of the good that can be done.

It is a dream worth clinging to, deeply.

The pectoral fin flickers but the trout stays still. Its head remains placid, implacable, hanging in the flow of the Test, its eye hewn and worked to a keenness, as if from flint. The tail fin adjusts, attuned to subtle differences in the current. Shark smiling, the gills open, revealing the pink slip of its capillary-rich filaments inside. Rock-filtered water filtered in turn through those gills, dissolved oxygen diffusing into its bloodstream, charging the trout with potential. It stays almost totally motionless, the many little flickers, tiny changes that keep a fish still in the flow, as apparently unchanging as this stream.

It could be dreaming in all this green.

A shovel-headed pike drifts gently, nuzzles the sky, passing itself off as a stump uprooted, just driftwood, not deadly danger. Banded demoiselles skitter through the pools of sky on the water's surface. A grayling passes like a bar of silver with a slack ship's sail for a dorsal fin, folded back on itself with the gentle lethargy of the afternoon.

Light drifts. Under trees my broken polarising sunglasses are useless. The water is clear but shaded dark. Glare comes and goes. The speckles on its flanks, dark spots in a beige halo, stand out as if polished by the water then blur with the speckled stones against the pale chalk bed, as the light and life and stream move around it.

Time passes.

Cramp. I shift. It shoots. A split-second reaction, zigzagging off, a golden lightning bolt disappearing into the dark of the river, the latent potential spent for that, my one false move. I never see it again. But it lives in this memory, indelible, inscrutable.

6

Under the Skin

At any point in the last ten thousand years, our droplet was picked up and drifted north and west. It splashes into wet ground. Moss drunk, it is trapped. As the moss dies, our droplet descends below the surface, swilling through the sodden layers of the bog, a solid-looking liquid, a liquid-behaving solid; the land preserved as if pickled by the rain. And there it lies, waiting.

Milly Revill Hayward, the RSPB's peatland engagement officer, plucks a leaf from a branch. She crushes it, sniffs it and hands it to me. Its leaves are strikingly like those of an olive tree; the same perfect droplet-shape, the same dull green that somehow shines. But they are perfect in miniature.

'Bog myrtle has got that lemongrass scent,' she says. 'It was once used as a midge repellent. Not that there's many of those about today.'

Milly pauses, glances up at the late afternoon sky; pewter grey and currently not raining for the first time in two days. I look across to nothing: visibility of a hundred metres or so and only heather, grass and bog myrtle, the land dappled with plants and lined with the scars of old peat workings. Bog myrtle is the tallest plant in the landscape here, but it grows no taller than knee height. No undulations, no relief, no breaks in the cloud that walks with us, shrouding the scenery.

As chalk streams are England's great contribution to the world's water, blanket bog is Scotland's. This particular one sprawls over a 4,000-square-kilometre swathe of Caithness and Sutherland in the far north of Scotland, the largest bog in Europe. It is called the Flow Country. The word comes from the Old Norse, *flōi*, meaning 'marshy ground', a reminder of the contested heritage that meant this corner of the Highlands wasn't brought under Scottish control until the thirteenth century. It's an extreme place, in habitat and climate: the combination of flora and fauna here is unlike anywhere else in the country. It's been around since the Ice Age yet it is fragile enough to be destroyed by a decision in a boardroom. We need to learn to love bogs – but it's not aways easy.

Milly fell in love with the Flow Country five years ago at a conference. Inspired, she took up a voluntary post here. But, as a Highlander, she confessed when we met earlier that there was a time when she did not appreciate the Flow Country, when she used to drive through and think there was nothing here. Which is an attitude I had too. When I used to take public transport to Orkney, the train or bus would trace their separate routes through the Caithness and east Sutherland peatbogs, the hinterland just back from the coastline in the very northeastern tip of Scotland. I was aware that the landscape out of the window was a rare habitat, a unique place. But all I could see was disappointing; a lot of empty space, as if mainland Scotland beyond the Highlands was where the hills had been worn out, ground down into an endless bleak nothing.

I am here now to look deeper: to see in detail what I missed from the train window, to see what this vast space of land and water has to offer. It begins with a boardwalk, taking us over the bog's fragile skin. Beneath the wooden slats is a growing, breathing, fragile living world where even a stray foot can cause damage.

Without the horizon to distract us we look more closely at our surroundings. Colour is muted: spring heather sprigs in green and grey; grasses, tousled by the weather, are olive and brown, slicked with rainwater. The caterpillar of a northern eggar moth clings to a blade of grass, the long hairs of its body jewelled with suspended water droplets. The white flowers of chickweed wintergreen droop down below their leaves as if depressed. This place is wet, evidently, from the weather. But this is a waterland where the real water is hidden; it takes a little knowledge to see it – and to understand what's so special about this particular bog that made me travel 480 kilometres to the north coast of Scotland instead of 8 kilometres down the road from my house. My local bog, Kirkconnel Flow, is a lowland raised bog – a beautiful curio, a bulge of peat and water surrounded by dry fields. But it is just a curio, an interruption to the landscape. At Forsinard, the RSPB's reserve in the Flow Country, the bog *is* the landscape.

The Flow Country is a land of paradoxes. Caithness and east Sutherland is the Highlands beyond most of the hills; an upland habitat, flat and at lowland altitude. This is where mainland Scotland tapers to its finishing point between north and east coasts. It is a waterland that appears to be mostly solid land but this is also a place where appearances wilfully deceive. It is called the Flow Country but it is mostly still. 'Reason is somehow unmade in bog,' writes Mark Cocker.[1] Time slips in the soft slick of peat.

Peat is the thing that makes this landscape special: it is dead plant matter that has only partly decomposed in the body of the bog, where the wet conditions prevent the oxygen required for decay. This means as well as being a special habitat for all manner of creatures, it is also a natural carbon sink – the 400 million tonnes locked up by the Flow Country is double the amount kept by Britain's woods. But, crucially, only if it is undisturbed. Like so many of our wetlands, however, blanket bogs have also suffered human interference, threatening the existence of the bogland, the creatures and plants that live here – and the release of all that carbon. And given that bogs take thousands of years to form, once they are destroyed, they are very difficult to get back.

About 400 million years ago, Lake Orcadie stretched from Orkney, across Caithness and east Sutherland, down to Inverness. It laid down flat sheets of mudstone, sandstone, siltstone. And then nothing much happened here until 10,000 years ago. The Ice Age passed, the glaciers scrubbing the ground, pulverising their till into the bedrock. It formed a dense layer of soil, like a clay, over the flagstone floor beneath, making it harder for water to drain through. Conditions were good but not quite right; the climate had wild fluctuations still. In the quickly warming climate after, grasses grew, then shrubs – dwarf birch, growing to shin height, and juniper – before trees colonised the land: birch at full height, oak and pine (and the chickweed wintergreen, a woodland flower that persists after the woods are lost). Then the early settlers of the Mesolithic, with their hand axes of stone, hacked down the trees, reopening the landscape. The climate cooled again. Trees no longer grew. Rain kept falling and, with nowhere to drain, the water pooled. The vegetation died and didn't decay. The now unfavourable conditions pushed the Mesolithic peoples from this land. But the stage was set for the bog to form.

The *Sphagnum* mosses are the makers of this world. When rain washes over them, they hold on to it – something like twenty times their weight in water, Milly says, as we peer beyond our feet.

By the dubh lochans* – Gaelic for 'black pools' – Milly finds three species. The first by fishing her hand into the water and bringing up a sopping wet string.

---

* Small dystrophic pools, acidic and peat-stained: found on blanket bog, so common in the Flow Country and the northwestern Highlands; rarely found elsewhere, where they drop the Gaelic name.

'So this is *Sphagnum cuspidatum.* Drowned kitten moss. It's supposed to look like a wet cat's tail.'

If you squint you can see it – if the cat in question was chlorophyll green and fitted into the palm of your hand. It is a limp string, swollen and shaggy with water like ruffled fur. Milly closes her hand. Water gushes out – an apparently impossible amount, as if this was the world's weirdest magic trick.

Next to the lochan, a hummock of red parts the green heather shoots.

'*Sphagnum capillifolium*. It's the same species as the green one,' Milly points to another hummock. The palate varies between apple green where they are shaded by others and red wine where they drink in the light and all the shades of pink-brown between. Each capitulum – the crowning shoot of the moss – looks like the damp tip of a fine brush, ready to paint watercolour across the bog. But bunched together they become like florets, like an eccentric bog broccoli.

'They prefer the drier ground. And this,' she picks up a miniature bouquet, two small bunches, 'is *Sphagnum papillosum*, it's chunkier still than *capillifolium*.' Chunkier and more ornately arranged, the capitulum is the centre of layers of branches that splay out like stars, stars of dull green in the bog firmament of watery greens and reds and browns.

A bog can be divided into two layers. The top layer, the skin, is called the acrotelm: from the surface it is 30 centimetres deep and it is the living section, where the plants grow. Under the skin is the catotelm, the dead layer. Not only does dead *Sphagnum* still hold on to water, but it also contains phenolic compounds, which are acidic. Combined with the waterlogging of the soil, they prevent the complete decay of the dead plant materials from happening. This plant matter, in a sort of suspended zombie state, becomes peat (and were it to be crushed under ground for many millions of years that peat would become coal). So under the living layer is a vast store of carbon and water. Healthy peat is 90 per cent water (there are fewer solids in peat than in milk). This is the end result of water and plants and acid – and time.

One of the things I love about bogs is that they are a landscape in four dimensions. Time is woven into the very essence of a bog, each year laid down in another millimetre, readable like the rings of a tree. A metre of peat records a millennium of rain. For the author Robin Crawford, peat suggests Ovid in its metamorphosis from plant to soil: he imagines 'the layers of peat like words and sentences on the page, forming stories of days and weeks; the paragraphs and chapters built up like banks of cut peat running across the moor to months and seasons, until the whole book, spread across the full year'.[2]

The story it tells is an epic. Slip your index finger into a peat bank and you'll touch Chernobyl between your second and third knuckles. Stick your hand completely in and you'll touch Tambora, the volcanic eruption that caused the 'year without a summer' and led to Mary Shelley's writing of *Frankenstein* and Byron's 'Darkness'. If you kept going up to your elbow, you'd touch the pollen that was laid down in the years of Shakespeare. Pollen, volcanic ash, even ritually sacrificed bodies have been found preserved in peat bogs. And if you work out how old you are in peat-depth time, you'll suddenly feel very young and very small; insignificant beside the bog of time.

There are 4 metres of peat beneath us here by the dubh lochans, which means that this particular part of the bog is as old as Stonehenge. Elsewhere in the Flow Country it has thickened, undisturbed over time, to 10 metres – a living, breathing relic of everything that has happened in that location since the Ice Age.

Across the world, 2 per cent of the earth is peat. Scotland is soggier and boggier than average: 20 per cent of it is peatland, and 13 per cent of that is blanket bog. To qualify as a bog in Scotland the peat must be 50 centimetres deep and the average bog here is between 1 and 3 metres.

Blanket bog is where peat gets to exist on a vast scale. An upland habitat, it is ombrotrophic, meaning it is fed by rain; rain that exceeds what the landscape loses by drainage, evaporation and the transpiration of plants. In Britain and Ireland they are skewed to the west and north, where the hills and rain and the cool climate are so ideal for their development that 20 per cent of the world's blanket bog is found here. The cool climate is key. *Sphagnum* doesn't do well over 15°C.* This is the pattern across the world, where blanket bog is pushed to the cold, wet edges: Scandinavia and Russia, Canada and Alaska, Patagonia and New Zealand.

There are other bogs: quaking bog, where a mat of peat-forming vegetation grows over a pond or other wetland, and standing on it makes the vegetation quiver and shake beneath your feet; and lowland raised bogs, where the quaking bog has grown over the water for long enough that the peat bulges below the skin of *Sphagnum* and forms a dome, raising the bog above the local terrain. But these are all small outcrops, strange jewels in the wider landscape.

The Flow Country extends across 200,000 hectares, with 21,000 hectares of that managed by the RSPB, but the wider area of blanket bog across Sutherland and Caithness reaches 400,000 hectares, which is too vast to get one's head around easily. London is 159,000 hectares in size: you could fit it into this area two and a half times and still have a substantial amount of peatbog left over. This is, I recognise, a ridiculous comparison. But this is a ridiculous space, a place where facts and statistics serve only to slip the mind's grasp.

---

* Although you can get peat in areas of the world that don't have *Sphagnum* or cool climates. Massive peatlands underlie the rainforests of the Congo and Indonesia where heavy rainfall and greenery create peat in the same way.

If these bogs are all in good condition, they do amazing work: breathing in carbon dioxide through their surface vegetation and breathing out oxygen. The inhaled carbon gets converted to cellulose, which forms the cell walls of the plant, until it dies. If it becomes peat, that carbon gets locked up in the body of the bog; if it lies undisturbed, if the peat remains wet and healthy, the carbon stays. Across the world's peatlands, that's 0.37 gigatonnes of carbon dioxide locked up a year.[*,3]

The RSPB visitor centre display offers this incredible fact: if all the Flow Country carbon escaped into the atmosphere, it would have the same climate impact as burning all the fossil fuels that Scotland uses for 100 years. I can't process that scale.

But, of course, it hasn't remained undisturbed.

We are fortunate to have the Flow Country. Fortunate because NatureScot estimates that 70 per cent of Scottish blanket bog and 90 per cent of its lowland raised bogs[†] to have been 'damaged to some degree'.[4] (I asked Milly if there was any bad peat bog locally: she said that their bad peat was what other reserves consider to be their good peat.) This is a landscape that has been taken for granted, unappreciated.

'Wetlands are wastelands; that, at least, is the traditional view.' So begins Edward Maltby's *Waterlogged Wealth*, a defence of wetlands published in 1986. He continues, 'Such apparent waste can only be put to good use if the wetlands are "reclaimed" for agriculture or building. Nothing could be further from the truth.'[5] Indeed – but

---

* A gigatonne is a billion tonnes.

† Like my local, Kirkconnel Flow. Lowland raised bogs are formed in places where drainage is impeded and the waterlogging prevents decay, creating peat. They then bulge above the ground level, hence the name.

that attitude has only recently begun to be righted. It's amazing we still have the Flow Country. Its legendary awkwardness hasn't stopped people from trying to find ways to destroy it.

Because peat is rich in carbon, it is combustible and fertile and where blanket bog forms there is generally a lack of trees. Crofters dug it – and still do on the Outer Hebrides – burning it to warm their houses. Distilleries malt their barley over peat fire to give their whisky the distinctive punchy smoke.* Glenmorangie used to cut their peat from Forsinard and the ridges you can see from the dubh lochan boardwalk are the evidence of their actions from the start of the twentieth century, where the nearby train station allowed the peats to be easily transported to their distillery at Tain.† The Scotch whisky industry says that its use of peat is less than 1 per cent of the peat extracted in Scotland per year. It is apparently committed to 'responsible extraction' and 'minimising impact' of its less than 1 per cent, though how this is possible is unclear to me.

Most peat in recent years has been extracted for horticulture, packaged into bags of compost for sale in garden centres and for growing plants in nurseries. There are moves ahead in Scotland to ban the sale of peat; in 2022 the UK government had announced it would do the same, but the plan was abandoned due to the 2024 General Election. Now it lives in a limbo, waiting on a private member's bill to push it through. In the meantime, the Royal Horticultural Society has committed to leading the way on peat-free gardening in the absence of governmental action.

Ireland has a strong tradition of cutting peat as fuel for domestic fires: 99 per cent of its raised bogs have been lost. The Irish National

---

* I love an Ardbeg or a Laphroaig but I'm not sure how this burning of peat can be justified any more. Offer me a dram and watch me have a crisis of ethics versus desire. I am fallible.

† Their whisky is now peat free and delicious.

Parks and Wildlife Service website talks in the wrong tense: 'The loss of Ireland's bogs would result in an irreplaceable loss to global biodiversity.'[6] I would suggest if 99 per cent has already been lost, that 'would result' should be 'has resulted'.

The Flows nearly suffered a similar fate. On my computer screen I watch Basil Baird, a farmer in a suit, drive his Rolls Royce Silver Shadow down the same single-track roads I have walked, past the pylons and across a small bridge over the River Halladale. He parks to a crescendo of flutes and violins and piano and emerges from the car in an ill-fitting grey suit. The frame wobbles, the screen flickers, the footage grainy, colours warping occasionally. It is an episode of BBC Scotland's *Landward*, broadcast sometime between 1977 and 1981* and uploaded to YouTube for posterity.[7]

The Baird brothers, according to the presenter, 'are planning to cash crop these peat bogs. This bold initiative, in one of the most undeveloped parts of the UK, could well revolutionise production from thousands of desolate acres like these.' (Never mind the peat that has been developing well here for thousands of years.) The Bairds were a pig-farming family from West Lothian. But having bought the Forsinard Estate, they had a plan to carry out their father's vision: that these bogs could be turned to hay. Cut to a montage of vast-wheeled tractors sliding over bog, double-wheeled tractors ploughing the peat, caterpillar-tracked machinery doing something unfathomable. 'Arrogant power,' says the voiceover, admiringly. Basil Baird says they're 'trying to make something out of nothing'.

The Ben Griams sit in the background behind a waving field of green grass. The programme descends into dull technicalities – farming talk for a farming audience – a focus on money in and money out, techniques for laying roads across peat (it's all about the

---

* The Roman numerals on the credits are indistinct, lost to the pixelation.

layers of rubble apparently), discussed over traumatic imagery: peat being cut and sprayed and ploughed; the habitat destroyed while people talk money. The last words are Basil Baird's: 'It's what the Highlands need, something for folk to do.'

By May 1981, the *Glasgow Herald* reported a different story: only twelve folk given something to do instead of the fifty to sixty that the Bairds had planned for.[8] Most of the estate up for sale except for the sown fields. James Baird blaming the economic climate; the price tag for the estate was £700,000 after they had sunk £1 million into setting up their farming operation.

Arrogant power will get you only so far here.

At the end of the boardwalk, the dubh lochans that pattern this landscape nestle into the heather; a series of silvery ponds, irregularly shaped, like the pieces of a jigsaw of water waiting to be fitted together. They are full of animal and insect life but that is not apparent right now. Dwarf birch, an upland shrub, grows here at low altitude due to the extremes of this place; its leaves like a miniature silver birch are the size of my little fingernail. It grows no higher than the top of my boots. It is a hint of the scale you need to look at this place, with a closer attention than elsewhere. This is a place, as Milly says, that rewards time and looking.

My favourite sentence ever written is J. A. Baker's 'The hardest thing of all to see is what is really there',[9] and the Flow Country proves the truth of that over and over again. Its landscape is charitably described by Cambridge University geologist Peter Friend as 'one of the less dramatic Scottish areas'.[10] It is in fact surrounded by drama in every direction, but to the point where it is easy to ignore what's happening here, as I used to. Even its defenders, such as the Flow Country Rivers Trust, know it is a thrawn place, difficult to love. Its website describes the area as 'mile after mile of rough, wet, awkward land'.[11] Derek Ratcliffe, its champion and saviour, wrote that 'I had never seen such a desolate landscape . . . there was a distinct similarity to the loneliness* of Arctic tundra.'[12]

Pedants will point out that it is unlike tundra, where rain is rare and the waterlogging comes from permafrost, but the comparison feels right. And the similarity is not just in loneliness. It is in the spareness, the stunted vegetation, the feeling of altitude at only 150 metres above sea level, the unforgiving emptiness. The climate and nature of these bogs is not like anything else in this country. As people here remark (as they do on Orkney and Shetland too), you are closer to the Arctic circle than to London. No need to look south. It is the north that explains this place.

The dubh lochans – known to science as patterned mire – are like the pools formed on tundra by the action of ice, the freezing and thawing that ruffles the surface of the ground. That doesn't happen here in the (relatively) mild climate of Caithness, where the reason for the formation of the dubh lochans remains a tantalising mystery: sometimes the bog resists the explanations of better-known landscapes. As late as 1990, when the rest of Britain was well surveyed and thoroughly known, Ratcliffe wrote, 'Such a wilderness offers the chance of making some new discovery. Much of it is still unknown and who knows what the next year will bring?'[13]

Whatever might still be awaiting discovery, there is plenty to appreciate that we do know about. There are another twenty-nine species of *Sphagnum* found across the Flow Country. Non-sphagnums

---

* It should be pointed out that this loneliness is the result of the Highland Clearances, when the Duke and Countess of Sutherland removed tenant farmers from their estate in favour of sheep. In the next strath west of here, Strathnaver, the property manager Patrick Sellar infamously burned down a house in 1814 while a bed-ridden woman lay inside. She died. He was cleared of culpable homicide at court and became a successful sheep farmer on the lands he cleared.

are obvious too, such as *Racomitrium lanuginosum*, the woolly fringe moss that forms prominent grey-green hummocks across the bog like rough ants' nests. Up close it is the shaggy twin of *Sphagnum capillifolium*: where that moss is made of a million neat paint-brushes, the brush-tips of woolly fringe moss are split and frizzy, as though they've been abused by a term of primary-school painting lessons. Its leaves are khaki but the frizzy hairs are grey, giving it a frosty look, fitting with the wet-slicked glistening of the bog. It does not form peat but it clings to water anyway. When starved of it, it slows itself down, as if sleeping out the dry season.* There are others that defeat me, which must be hiding in plain sight in the botanical and bryological carpet of the bog; trying to pick out just one thing here is like looking at a clear night sky and trying to keep track of an individual star. This is a place where the individual blurs into the collective.

Nothing buffers the acidity of the bog. It means that mostly specialist plants grow here. Great sundew stands up from the wet patches between grass and heather, reaching up with a tea-spoon-shaped leaf – though leaf seems the wrong word for it: growing out of its green are red filaments, thicker than hairs, with a bulbous end like a droplet of blood, unlike any leaf I've ever seen. Clear spheres of a sticky, sugary mucilage are suspended in these filaments. When an insect, attracted by the sugars, lands on the sundew's leaf, it is suddenly stuck. The leaf rolls up and the insect is slowly digested by the enzymes. Round-leaved sundew does the same thing but its circular leaves are held out from a central stem like a warped and weird clock face, just above the *Sphagnum*. Butterwort flowers, like purple mouths held agape, reach out to the height of the heather

---

* Which weirdly reminds me of the globe skimmer dragonfly and its migration on the continent-crossing rain fronts of the tropics, giving it the appearance of having materialised with the rain.

shoots.[*] Their leaves, five-pointed stars like a lime-green starfish, have a greasy feeling that also traps and digests insects. This is a place where life is hard, where the need for nutrients turns plants into predators.

The RSPB has a lookout tower here, a sheltered spiral staircase that leads up two storeys to a platform overlooking the dubh lochans. It faces towards the Ben Griams (Beg and Mor), two distinctive mountains that lurk in the background of every photo I've seen of the reserve here. Except for today. Visibility ends a little way beyond the lochans, barely a hundred metres. A swallow flies in circles around the tower, desperate for the insects that aren't flying today. This place is vast; on a clear day it can make you feel small, totally out of scale with the surroundings, but this evening it is claustrophobic, resisting our intentions and attentions. Even the hare's-tail cottongrass has had enough: its signature seedhead, a single ball of cottony fluff like a scut, has been slicked by the wet into a soggy grey droplet.

Talking it through with Milly, she slips into poetry.

'I love it when it is like this. There's two blankets. One of the peat below and the other the mist above.'

And in the thin seam of light between the two, they are joined by the lochans; the water, the defining feature of this landscape, above and below. Even if most of it is under the skin of what you can see.

Rain begins to fall again. The light begins to fail.

Dawn happened in gradients of grey. The morning waking up in a hangover of last night's rain; a mist hanging low enough to

---

[*] These flowers happen to look a lot like the Rolling Stones lips logo in purple.

stifle the light and cover the view, everything glazed with water. A few kilometres north of Forsinard is Forsinain, an RSPB-managed farm in the middle of the peatland reserve, an island of grass. A trail runs through it to the heart of the bog from the edge of the River Halladale, its water running like a pint of Guinness being poured; black with peat, the rippling, bubbling flow of it headed with cream; a common sandpiper calls and flickers across its surface, house martins and swallows swirling, chasing insects. Midges are coming through in waves. Not the biting kind* but the irritating kind, those that seek out the corners of eyes, nostrils, the insides of ears.

The silence is almost total. There is no traffic on the road. Just the soft rustling of grasses in the breeze; the chattering of swallows and the tinnitus of midges. Over the hillside three snipe are displaying. They are pot-bellied, lined in browns and blacks, a deeply earthy species until they display when they fly wavering lines through the sky. Their outer tail feathers are held apart, out at a right-angle between the tail and the body. As the air passes over that feather it vibrates, casting a sound; a whinnying sound but this undersells it, an unsung song that also sounds deeply uncertain, like the snipe is umming and aahing through the sky.

At the top of the slope is the rough ground of Forsinain Hill. The farmed land suddenly stops, abruptly, with a wall. The mist had been receding from the strath floor but it thickens over the damper ground of the bog. Waymarkers point the way to a flagstone path, bunches of *Juncus* offer secure footing between the softness of *Sphagnum*. By the flagstones I realise with horror that the wellies I'd brought along for this occasion are still sitting in a bag in

---

* They could be chironomid non-biting midges or, as biting midges are females requiring the protein in your blood to enable them to breed, they could just not be in the mood at 7 a.m.

the car boot and that the path has sunk below the water table. No going back.

Peaty water pools around my toes, soaking into my socks. It is not the sudden biting cold of the Clyde but a gradual growing dampness. I rock back onto my heels and wobble precariously over the worst parts. Some of the flagstones slip on their uncertain ground, underfoot, air bubbling through the water below like a whoopee cushion. A northern marsh orchid grows from a hummock between flagstones – on the coast they are flowering now but here, mistier and wetter and cooler, it is a rosette of speckled leaves and a short stem, crowned by a point of buds, ready to erupt in magenta.

Dubh lochans open up again. They have been refreshed by the recent rain but through the clear water I can see dried peat, cracked like broken glass, a reminder of a recent dry spell. Bogbean is prominent here, growing out of the *cuspidatum*-lined pools, their stems a stubble, doubled by their reflection. Its green leaves hang like flags that can't agree on the direction of the wind. Its buds are broad-bean sized and shaped (hence the name), erupting in turn up the stem, from base to tip, in white stars. It is a strange plant of the waterlands, particularly of these peatier, acidic, harder-to-thrive-in places. It is monotypic, meaning that it is in a family of one, its evolutionary branch ending with it alone and no close relations, as singular as its habitat.

The mist makes birds invisible but sharpens their calls, tunes them to unsettling. I can hear the shrill eruptions of a singing dunlin, the Hammer-Horror hauntings of curlew, the vocal cords of golden plover like an aeolian harp, whispering the wind through taut strings. Unerringly there is the sense that they know I'm here, that they have seen me coming, recognised that I am out of keeping with this landscape.

The mist lifts a little. Along the hillside I can see a shattered plantation, trees as stumps, the brash laid into the ground, the first steps in the restoration process of forest to bog. Beyond I can see the mist clinging to the pines of the next forest to restore. But I can't continue. A sign set across the path reads, 'Trail closed to protect breeding birds.' I'm happy to give it to them.

I double back. Another track leads to a viewpoint over the bog. A loch is just visible, a thin strip of water, pale as bone in the dark heather. There is initially nothing to be seen. Some landscapes reward watching nothing, waiting to glimpse a movement. Some landscapes, such as woodland, reward not looking at all but listening. Bogs reward both active searching and listening, panning with the telescope side to side like a prospector.

I strike silver. Standing vigilant on a hummock, not breaking the skyline, a greenshank. Like its close relative, the redshank, it is an alert wader, beady-eyed, constantly looking around, while its feathers, a gradient of white, silver and grey, stand out starkly against the dark vegetation. It stalks off quickly, all hyperactive, jerky movements. A minute later and it whistles noisily, three fluted notes. From cover it detaches up into the sky with two more unseen greenshanks, chasing them off its patch of bog. I last saw one on the Clyde in winter. Here it feels like a completely different bird. There it was elegant, peacefully sleeping. Now it is vigilant. Its relative, the redshank, is known as the warden of the marsh. The greenshank isn't known as the warden of the bog – their southernmost extent as a breeding bird are these lesser-trodden corners of the far north of Scotland where they carry an air of mystery about them – but it should be. Its strident calls echo around the empty spaces, ever alert and warning of anything. I slip away before they can alarm about me.

Turning around, I find a place where the skin of the bog has been ruptured. It is most evident by the side of the roads or tracks where the bog skin has been cut. Under the overhanging heather, water drips down from the side of the peat. Elsewhere moving water cuts gullies into the bog; from higher ground you can see lines like contours where burns have creased the peat and worn away at the vegetation. Where this happens, hags (from the old Norse *hǫgg* for 'gap') form, isolated islands of peat, bare-sided and surrounded by gullies. From here, peat exposed to the air can oxidise and dry out; releasing carbon, turning to dust and blowing away. Deer have a similar effect where they stampede over the same soft areas, or where they have grazed the vegetation too short.

I push my finger through the exposed, drying black bank by the track. Peat is often described as being like a fudgy chocolate cake. I understand that. The texture of this drying peat is like a cake that was once moist; its dense stale crumb held together by the remaining dampness of the earth. This peat isn't particularly healthy – it's definitely not 90 per cent water – but it's not dead yet.

I return down past a different farmed field of grass – I have to head back to the hotel to change my socks and pick up a bottle of water.

The hotel has a private water supply, so the water that comes through the taps is not from the mains. A vague warning poster says it comes from the 'Local Water Course', which could be anything here, where the building is an intrusion in the middle of a giant watercourse. Multiple signs advise that despite the 'minor discolouration and potential small particles' it is safe for washing in but not for drinking. I don't test that. But I do pour a glass of it and it comes out tinged golden, like diluted whisky. I wash my flask in it instead and the next coffee has a tangible earthy note. Surrounded by all this water that we aren't advised to drink is mildly ironic. Against my better principles I have to rely on the plastic bottles of Nestlé water the hotel provides.* (I take a lot of recycling home.)

---

* In 2005 the then CEO of Nestlé said that the idea of water as a human right was 'extreme': https://www.snopes.com/fact-check/nestle-ceo-water-not-human-right/

Water that comes from either Derbyshire or Pembrokeshire and shipped around the country before coming here, where the water is already absurdly plentiful.

I take a sip. Out of the window a block of conifer plantation stands, awaiting its restoration to bog.

After the Bairds' failed attempt to transform the bog into farmland, the estate was taken on by a forestry operation.

First Galloway disappeared under conifers, then the drive to plant moved north, following the apparently empty lands; the industry incrementally working out how to plant on deeper peat, more intractable land. It was not just the work of the Forestry Commission but private companies too. They were given generous grants and tax breaks to assist with the process of establishing their plantations. Because this was the 1980s, the system was baggy with loopholes. The tax breaks were supposed to be recouped by the taxable profit from the sale of the timber but by selling the half-grown plantations, the companies involved avoided the tax. In reportage for the *Observer* magazine in 1988, two journalists worked out that 'some end up with a return rate of up to 33.5 per cent a year on their original investment – which means that the investment has almost doubled in value in two years'.[14]

The ability to offset tax made this a highly desirable investment for the already wealthy, many of whom became subsidised forest owners in the Flow Country. If water flows to money, the peatland equivalent was that water grew money. In the 1980s a tenth of British peatlands had been forested with conifers;[15] 17 per cent of the Flow Country had succumbed.

To plant a pine on a peat bog requires preparation. The bog needs to be deep ploughed, the other vegetation discarded, and drained; the rich and fertile peat needs to be dry enough not to drown a tree. Aerial photographs from the 1980s show land covered in dense straight lines, narrowly spaced – unnaturally linear in this landscape of loosely defined edges and freeform shapes. Exposed to the air, the carbon locked up in the peat oxidises and is freed. Exposed to the air, peat stops forming, stops holding on to new carbon dioxide. The trees that grow are supercharged by the peat. They then further dry out the bog, sucking up the water for their own uses. A bog is a single hydrological entity: that is, planting trees on one part of it does not just affect that part but the bog as a whole. Having taken thousands of years to form, they can be undone more quickly than they can be fixed.

What made the Flows special was hit by this. The water-loving plants were parched by the water being drained. The birds of open space were put off by the trees that provided cover for their predators, isolating them in islands of bog surrounded by forest. The carbon absorbed by the growing pines does not equal that which is released from damaged peat. Afforestation was quick – there was money to be made.

Ecologists from the Nature Conservation Council, who were carrying out a five-year long survey of the peatlands, were 'racing against the forestry ploughs'. With dark humour they called them 'fire-brigade surveys'.[16] The surveys were uncovering what was specifically special about the flows, what the case for preservation could be built around, what people didn't know because they hadn't been looking closely enough at the landscape and the beauty of the bog. The forestry was initially popular – it created jobs in a deprived area and intervening was politically tricky, especially as it was one arm of the government essentially criticising another. The NCC published a report criticising the afforestation of the flow country and the local MP retorted that 'the NCC seemed bent on "sterilising" the land, just like the Highland Clearances of evil memory'.[17] The NCC won the argument – a long, complicated argument pitting ecology against economics – and the then chancellor Nigel Lawson stopped the tax breaks in 1988. The Forestry Commission updated their regulations about planting on peatbog.

Derek Ratcliffe, as the NCC's chief scientist, was instrumental in its campaign to save the Flow Country from afforestation. He marshalled the data, stress-tested the science and lobbied the NCC chair to fight the battle but his contribution was almost spectral; he was always reluctant to take credit. In a 1984 publication he argued that 'posterity will judge us by deeds not words'.[18] On 8 July 1989 he retired. And three days later the NCC was dismantled, its victory pyrrhic, split into three regional bodies of reduced power; the revenge of an embarrassed government.

The trees kept growing. By 1995 the RSPB had raised enough money to buy the Forsinard estate. Over time their land holdings increased until it became their largest nature reserve, fuelled by the donations of its membership, who recognised how close to disaster it had come. By providing a boardwalk and the trails at Forsinain, they protect the bog from uncontrolled disturbance, while the reserve offers a great laboratory for peatland restoration: blocking ditches, returning water to the land and removing the worst of the pines.

The trees, meanwhile, are about mature now. Two million tonnes of timber will be produced out of the remaining Flow Country forests by 2030. After all of that destruction, the timber will be poor quality: water often finds its way back to the drained sites, slowing the tree growth; trunks grow crooked; pests proliferate; the wind has toppled others too soon. Most of what grows gets pulped for paper or wood chippings; only 20 per cent gets turned into products such as pallets or fence posts. Those forests in the areas of the bog that are being actively restored are felled and left to rot; returned to the earth to begin that journey of arrested decay into peat.

But in turn another potential threat pops up. The flat exposed Flow Country is the perfect place for wind turbines – Caithness and Sutherland had 380 turbines by 2020 with plans approved for 165 more. But any disturbance to the peatlands risks emitting carbon, making a not-so-green renewable energy source. The industry talks well about peatland restoration but the risks are various of disturbing the peat, altering the hydrology of the bog, pollutants unintentionally released. A more practical concern is the map on the Highland Council's website of wind turbines that have been built, are being built, are planned or have been refused. There is a clear bias to the edges of the Flow Country, particularly along the coasts. Among the unique birdlife of these bogs are breeding divers, skuas and sea ducks including common scoter that commute from the lochs and lochans on the bogs to the still rich seas between Caithness and Orkney. Lining their routes with potentially lethal wind turbines is like asking them to run a gauntlet every time they forage at sea.

The Flow Country is used to fighting for itself. But in 2017 it began to strike a pre-emptory blow. The Flow Country Partnership began working on a bid to be recognised as a UNESCO World Heritage Site: international recognition for a bog, as important as Stonehenge or Skara Brae; Angkor Wat or the Taj Mahal. Their justification was that nowhere else in the world has the array of species found here, neither for its combination of Arctic, Alpine and maritime birds or in its Atlantic and Arctic flower assemblage. In 2024, the bid was successful – it is now the first peatland World Heritage Site. Which is extraordinary. UNESCO explicitly states that the purpose of World Heritage Sites, whether cultural or natural, is that they are our 'legacy from the past . . . and what we pass on to future generations' and that they 'belong to all the peoples of the world'.[19] Peatlands, the time-keepers of the nature, in their slow development and climate-altering effect, seem self-evidently to fulfil those criteria to me.

*

The bogs may border on the monotonous. Hydrologically they are (or are supposed to be) one entity. But the ones I have described so far have been the West Halladale peatlands. Thirty kilometres east of Forsinard as the eagle drifts (or an hour and a quarter by car) is another nature reserve, leased from the local estate by the charity Plantlife. Munsary Cott is not the most welcoming of reserves. Behind a rusty gate is a noticeboard that proudly trumpets itself as 'The Road to Nowhere'. To paraphrase: 'You are welcome to walk the 10-kilometre return trip down the track (to nowhere) but you can also turn around whenever. It is extremely difficult and can be dangerous walking and if you wish to actually see Flow Country plants, visit the RSPB reserve instead. Sorry, there are no toilets.'

Ahead of us a single angler works Loch Stemster from the far bank. Dunlin scurry along the near bank, disappearing behind stones. A snipe rockets from the undergrowth. Greenshank incessantly shriek, lapwings chase crows, curlews descend into thickets of common cottongrass, their multiple white pompoms shaking joyfully in the breeze, a blizzard that blows but goes nowhere.

This is a vast land. The grey miles of Caithness, across which I can see back to the faint teeth of Ben Griam Beg and Ben Griam Mor of West Halladale; the sort of place where 30 kilometres of land just falls away into the nothingness of space, a meaningless distance.

I walk the 5 kilometres to the edge of the peatlands, through the farmer's gate. The air beyond is grey, thick with evaporation or the drifting distant smirr. Blocks of conifers stand dark and implacable like headlands sticking out in a sea of heather and *Sphagnum*. A huge bird circles but slips identification, a sense of size but no scale in the distance. A bull circles its herd and walks in my direction, on quicker steps than I'm comfortable with and a questionable curiosity about its manner.

Bogs have ways of resisting being known: their impenetrable hearts do not welcome the human; details of the billions of plants and mosses dazzle up close and blur to brown and green and grey across the scale of the view towards the horizon. Satellites are helping us understand bogs. Interferometric Synthetic Aperture Radar – InSAR for short – is a method of tracking the planet's surface through the changes in the time it takes for the satellite to receive the signal bouncing back from Earth. It is precise enough to measure the breathing of a bog, the expanding and contracting of its surface layer as the levels of water and gas in fluctuate in response to weather and the management of the bog. Research by Professor Roxane Andersen of the University of Highlands and Islands has mapped this against the risk of bog burst – where after heavy rain, the skin of peat is ripped open by the pressure of the water below, and a torrent of peat slides downslope. The analysis showed that the more active the bog surface – the bigger the fluctuations between sudden wet and dry conditions – the higher the likelihood of a bog burst and a sudden flow of peat into rivers and lochs, killing fish, releasing carbon, endangering the lives and livelihoods of any body or building in the way.

Back at Forsinard, I cross the railway. A hind stirs, stares at me from the northbound line and returns to the grass growing through the stones. It's 7.15 p.m. The occasional truck rumbles down the road through the long evening light. The birds are quiet still but there's two and a half hours yet until sunset. After half a day of dry sunlight, the hare's-tail cottongrass has dried out, buffed by the breeze until fluffy again; the caterpillars have shed their water jewellery and crawl like babies across the boardwalk; below, the chickweed wintergreen flowers turn to face the light again, their petals a circle of seven overlapping white teardrops.

In May 2019 the reserve here was touched by fire. It began on the north coast near Strathy and for six days it burned 57 square kilometres of West Halladale peatbog between there and the edge of the Forsinard (according to the RSPB it burned 3 per cent of the reserve). It was, at the time, the largest wildfire on record in Britain, and for the duration of its blaze it doubled Scotland's carbon emissions. Its smoke was visible from space. The smouldering, blackened vegetation left in its wake contained the cremated remains of bird nests, the bodies of lizards.

Peatlands are particularly vulnerable as they are profoundly flammable, which is kept in check by the amount of water locked up inside them (peat cut for a domestic fire has to be seasoned like a log). Degraded bogs, which are less wet, burn better.

The fire was fuelled by a hot, dry spring. But it was building on the drought of 2018 when the Flow Country received less than 75 per cent of its annual rain. Fairlie Kirkpatrick Baird, an analyst for NatureScot, built a model extrapolating trends to predict the future rate of drought in Scotland. Between 1981 and 2001 they found evidence of one extreme drought. Their model shows that by 2040 there will be an extreme drought every three years.[20] The basic west–east, wet–dry dynamic sharpening to extremes. And with it, *Sphagnum* becomes replaced by species that hold less water. The special birds of the bog evaporate. The bog becomes moor.

At 3.30 a.m. I stumble into the hotel corridor. Argue with a door. Pull a boot on the wrong foot. Condensation on the windowpanes. The diffuse pre-dawn light beyond. I'll regret this in a few hours' time.

I step outside into the fresh air. Breathe it in. I have no aim to being here now, I just want to steep myself in the bog one final time before I leave. A hind stands in silhouette in the gap of an open gate. Behind her a bank of mist sits in the strath, where the ground-warmed air condenses under the cold clear sky. That sky: three bands of lilac and peach and a pale blue grey that darkens by gradients, decorated by high wispy clouds that have become filaments of neon pink. The silence, extraordinary here, is intensified by the early hour. The sound of my feet crunching on the gravel of the road edge is amplified by the absence of any other human sound. There are two cuckoos echoing from the hollow plantations on each side of the road. A skylark and a meadow pipit and a snipe. Water trickles somewhere. The railway is quiet until the first train of the day passes through at 7.27. I step onto the boardwalk before even the cottongrass awakes.

Dew glistens. A snipe shoots off at my approach. Shortly after a greenshank wakes and slips more gently into flight. A red grouse hacks like a smoker's cough on waking; then another, and another, until I am surrounded briefly by avian spluttering; then, lungs started, they suddenly cease, disappearing from the hummocks into the heather. The mist creeps up the strath, building, a rolling tidal wave of cloud, dimming the dawn light. To the east the slow flow of mist out of the higher ground works like a river in the morning air, seeking out the lowest ground in the land to glide along, smoothly, stately, before drifting across the miles to Ben Griam Beg. It wears the mist like a scarf, clouds like a hat and blushes in the middle with the rosy-fingered light, just breaking through. The dubh lochans reflect the pink light and white clouds. To the north the mist is a dissipating fuzz, blurring the land and light.

I find myself walking not waiting. While the light is still slant, the bog is filled with shadows. Butterwort is the brightest thing, its lilac incandescent, jewelled in water; heath milkwort slighter splashes of a darker purple deep from the undergrowth. I find my way back to the road and walk to the south. The light, rising, partially breaks through and the Ben Griams are suddenly growing green between sharp black shadows. A lizard scuttles through the roadside grass, disappearing by a growth of lousewort. Their pinky-mauve flowers have five petals, the top two fused together like hands in prayer above three petals spread out in a semi-circle like an offering. Unlike butterwort and sundew, lousewort isn't predatory but semi-parasitic, its roots tendrilling down, tapping the nutrients collected by others to supplement what it can get by itself.

At the top of the road I stop. 7 a.m. The light won't get better. The straight road runs down the side of the strath before me, the landscape reopening to the east with four lochs named on the map, and multiple unnamed lochans. Mist lingers there. With the bright sun behind me it burns white and for a second, it seems hard to tell what's down there; as if the pylons fade out into watery paradise; as if there is nothing there remaining of human hand; as if this is a place where everything flows.

Water, acid, plants and time all build and shape the bog; that thing that defines and explains this corner of Scotland. Battered by our changing priorities – pines and dreams of hay and money to be made – the prospect of changing water is another note of alarm. The bogs of the future bursting or burning. But these are seductive places, whether shrouded by evening rain or dazzled by morning light; it's as if, like lousewort, the roots of the bog have tapped into me.

Bogs, like our rivers, have been in worse positions than they find themselves in now. And although the timescales of a river and a bog are vastly different, those bogs that are depleted, degraded or drained can – must – return to a wetter state. Or we let this – Scotland's greatest contribution to the world's waterlands – slip through our fingers, in a haze of smoke and carbon.

The tips of the long grasses hang together, weighed down by the dull, dewy morning. I part them, feeling fleetingly like a Moses of the world in miniature. In the space between, a leaf stands proud, the highest point of its world in hiding. A great sundew. It is like a strange star, a warped Jupiter in green above everything else, with red rays radiating around it. They seem to be a source of brightness as if they are the filaments of some surreal lightbulb, glowing in the gloom that surrounds it, lighting up this corner of the bog. That light is caught in the moons of mucus, the 'dew', a false water that rings the leaf, as if pulled in orbit. They are of a perfect clarity, reflecting the grasses and stems around them, innocent like the light of an angler fish until the seduction works.

It is the wrong morning. Too cold, too wet, for their insect prey to be blundering about between the grasses. It is frequently the wrong morning, in this rain-washed world, where every nutrient falls from the sky in a droplet of water, where there is not quite enough to go around. Later, when it brightens, a fly will blunder through, pulled in by the sickly light of a sundew's moons. It'll land on the leaf and be rolled up. The leaf's enzymes dissolve the insect, scavenging the nitrogen from its body, which the sundew requires for photosynthesis and protein.

All this on a whisker of a stem, curving like a curlicue as it descends into the peaty water, darkened by my shadow, where it disappears into a slumped mass of vegetation, mixed in with the mosses. Drowned but alive.

Wet-kneed, bent-necked, back-ached. I let the grass go. Out of the water here what grows is bloodthirsty. Land that traps life. Life here is like each step on the bog: unsettling, uncertain, never completely what it might seem.

7

Rebirth

The land sinks low. The droplet percolates. It swirls through the soil to the clay beneath, then rises again with the groundwater to become surface water; a cocktail of rainwater, river water, aquifer water. Enriched from the descent through the soil, the droplet drifts through reed and sedge, bringing nutrients to what grows, liquid to a dry place.

The marsh harrier lifts up from the reeds, its wings held in a shallow V, barely flapping, ascending by an effortless circling until, high against a backdrop of bare poplars, it breaks free into the sky. A sky of November light; pale blue above with drifts of snow-white clouds, not illuminated by the neutral afternoon sun. In the breeze the *Phragmites* reeds sway in waves, the golden seedheads the richest colour in the fen. With each bend the reeds shed their seeds and the air carries them, a barely visible dust, to settle in the edge of the water. Bulrush lines the open channel, straighter and stiffer, their seeds stored out of the wind's reach. Another marsh harrier drifts without intent along the river wall beyond the reeds. This fen is one of those places where, no matter what else is present, my eye will always be taken by that shallow V of wings – like a child's drawing of a bird in flight – quartering the reeds as if always on the verge of stalling, idling in the air, as if flight simply required no effort at all. It helps that here the viewpoint is so low and the

Phragmites so high that there seems to be nothing but birds between the veil of reeds and the distant trees.

A marsh is the classic wetland. It is technically defined by vegetation but it is also a place where water defines the land, soaking into it. It can be permanently covered by water or rinsed and repeated by the tides. The main division within marshes lies within the water. In a freshmarsh, freshwater soaks the land; in a saltmarsh (which we'll explore in Chapter 9) it is the sea.*

Freshmarsh is dominated by reed and sedge, soft-stemmed vegetation that can grow denser than trees. A swamp is a freshmarsh where the dominant vegetation is woody: in Britain, where we don't have great expanses of this, we call it carr; in Europe we find swamp mostly in Polesia, the borderlands between Ukraine and Belarus. Most of the world's swamps are in the Americas, Africa and Southeast Asia.

The finer differences between forms of freshmarsh are found in the water as well. A fen is a peaty marsh – where water comes from the ground or a river and the vegetation doesn't decay properly in the waterlogged conditions so it forms as peat.† In the north of England these are also known as mires. A freshmarsh that is not a fen relies on rain for its water – known as ombrotrophic – and lies over the top of non-peaty soil, and so freshmarshes are a bit less specialist and more frequently found. But these differences are pedantic and not particularly apparent to the non-ecologist. The species and functions can be the same. Some wetlands, such as the Norfolk Broads, combine fen and freshmarsh.

* With the exception of a few flooded old salt mines in the Midlands and northwest England.

† As opposed to bogs, which get their water from the rain rather than the ground and thus have more acidity. It is really visible in the plants: fens don't have heather; bogs don't grow reeds.

The Broads are lovely but the East Anglian Fens, adjacent to where I grew up, mean more to me. Such as this fen. The RSPB's Lakenheath Fen is a very specific middle of nowhere, near to where the edges of Suffolk, Norfolk and Cambridgeshire dissolve vaguely into each other. Its golden ribbons of reeds are flanked by carrot fields and matchstick plantations; military airbases and scrap yards. Trains rattle alongside the reserve and F15 fighter jets scratch lines of water vapour through the sky above. It is not entirely peaceful.

Some people think the Fens are not worthy of looking at because they are flat, as if the vast and changing sky held no interest, as if in the corners and edges interest couldn't be found. Their loss. This is not junkland. This is the land of extreme contrasts: where history has writ a place of dry straight edges and wet curves; green fields and threads of gold; where millions of these millimetre-long threads drift in a gentle breeze below birds that were once rare. Where oases of water can be found in the driest English landscape (which was once all water anyway). Where people may see all of this or nothing at all. Paradise or a problem. Because this landscape has been an intractable human problem, ever since people saw this amphibious, difficult land and imagined fields instead.

It has happened all over the world. Marsh and fen, drained to make 'useful' land – agricultural land – and in the process giving us issues: flooding, the release of carbon emissions, a land being erased before our eyes. In the Fens it happened to an extreme degree. But it has not yet been damaged beyond salvation; the harm can still be undone. This very reserve is a shining light in the world of restoration: a solution to the issues, the habitat rekindled; a place where life has returned in abundance.

After the Ice Age all of this was sandy, gravelly land, a vast dried riverbed where glacial ice once flowed; where pine and birch and willow were scattered instead of crowfoot and lily pads. When the sea level rose up and reclaimed Doggerland,* it also worked its way up the rivers and onto the low-lying land surrounding what we now call the Wash. As the sea ebbed and flowed, it left clays and silts covering the original vegetation and slowing down the rivers. The freshwater backed up, its way blocked, and began flowing over its banks into the land on each side. The land is so low and flat here that water, out of the flow of the main river, is sluggish, without gravity to power it. So it soaked down into the soil, until waterlogged, until the dying plant matter could no longer decay properly and it became peat. Repeatedly the sea flooded the entire fenland area over thousands of years, depositing its sediments in the rivers, its saltwater killing plant life, contributing to the growth of the peat beneath and making it desirable as farmland, where living plants are supercharged by the richness of dead plants, like garden-centre compost on a landscape scale. Ten thousand years of carbon that would otherwise have been emitted from dying plants became locked up under water.

This happened over the 4,000 square kilometres of eastern England that stretches between Lincoln and King's Lynn in the north, down via the northwest tip of Suffolk, Cambridge, and across to Peterborough; a shallow hole in England. A quarter of all the lowland peat in England and Wales was laid down here. It became known as the Fens and between the disappearance of Doggerland and the seventeenth century, it existed as a

* Doggerland was the land bridge between eastern England and what are now the Netherlands, Germany and Denmark, where bottom-trawling fishing boats bring up arrowpoints and the bones of long-extinct animals. Between the Ice Age and inundation, it is hypothesised that it would have been among the richest lands in Europe for hunter-gatherers: muddy and wet and full of wildlife. It had wetlands, of course, but now, drowned by the North Sea, they are a bit too wet to count.

wetland on a vast scale, comparable to the Danube Delta of the Romanian–Ukrainian border.

The largest lake of England's lowlands, Whittlesey Mere, a vast sheet of shallow water best measured by the mile, was here. As were swallowtail and large copper butterflies; wolf, bear, boar and beaver; white-tailed eagle and Dalmatian pelican. Megafauna waded through these wetlands, which are hard to reimagine from their current state, under flat farms and where a roe deer or swan might be the largest animal for miles. People lived here too. At Flag Fen and Must Farm in the heart of the fens, oak-timber trackways (with preserved evidence of Bronze Age axe cuts, some with beaver bite marks) have been discovered sunk into the waterlogged soil. Flag Fen seems to have been the ceremonial landscape: a place for burials and offerings, whereas Must Farm has revealed round-houses and logboats, fish traps, pottery and even a wheel – the oldest wheel found in Britain. It is reminiscent of the Scottish and Irish crannogs: life was lived suspended over the water, alive to what it offered.

In the 1970s Francis Pryor, the archaeologist who pioneered English fen archaeology, discovered Bronze Age field systems: what had been thought to be tracks crossing the fen were actually ditches that had filled up with silt and gravel; field boundaries that would hold water away from the summer grazing fields; drafting races and droveways. The inhabitants and neighbours of Flag Fen and Must Farm were not just subsistence farmers but had turned the wild fen to a seasonal, reasonably intensive farmland. The rich soil beneath the fens has been irresistible for millennia.

But the fen had been only partly drained and tamed by local people, because in Pryor's words they 'understood only too well how water behaved in these intricate landscapes because if they got it wrong . . . people might drown and family homes and farmyards would have to be abandoned'.[1] Vast wetlands still existed. The drained areas were common lands. Nobody's property and every local's shared interest.

That all changed in the seventeenth century when Cornelius Vermuyden, a drainage expert from the Netherlands, severed the link between the local people and their watery land. He was a tool of the Adventurers, a group of wealthy people who ventured their money to gain dry land where there was once only fen. Vermuyden directed the creation of a new river from 1630, a new route for the Great Ouse,* England's fifth-longest river. He wanted to guide the water along a more rational course: a straight line cut down a steeper gradient towards the sea, running through the heart of the fenland. In principle it would be like pulling out the plug on a sink full of water. But as Mr Crick says in Graham Swift's *Waterland*,† 'silt obstructs as it builds; unmakes as it makes'.[2] Vermuyden, who wasn't a charlatan as such, although it is notable that his finished schemes in Britain never quite fulfilled their aims, did not account for the nature of water, the nature of rivers and the nature of land. A fenland spin on the chaos theory of the Clyde: if you alter a river, the silt will alter it further than you can think. The Great Ouse kept flooding. The new river, named after Vermuyden's benefactor, the Earl of Bedford, had filled up with silt.

The Old Bedford River now sits parallel to the New Bedford River, another Vermuyden waterway cut along the same lines in 1649. While the purpose of these rivers was to drain the land, to stop it from flooding, between the Old and New Bedford Rivers runs a large swathe of washland – grassland designed to be washed over in winter with the excess water that neither of the rivers can

---

* The etymology of 'Ouse' comes from the Celtic for 'water'.

† One of my favourite novels. That the truest representation of fenland history is to be found in a work of fiction is one of the delicious ironies of this landscape.

cope with – a tacit admission of the failure of ever fully draining the fens. This land, 30 kilometres long and a kilometre wide, can hold 90 million cubic metres of water. In the snowmelt floods of March 1947, it did. With a certain irony, there is now a string of nature reserves running along this washland, where waders including the black-tailed godwit breed in summer and whooper swans spend the winter. Scraps of wildlife surviving in the scar tissue of this damaged place.

Vermuyden was not popular. When his works were being planned, an anonymous ballad began to be circulated. The 'Powte's Complaint' is voiced by a fenland fish, speaking to its 'brethren of the water'. It is a militant poem, playing on the fear of the incomer, the removal of the land from the local: the third and fourth stanzas each end with a vision of the wetland drained, killed because 'Essex calves want pasture'. Throughout it contrasts the local way of life with the threat to come: 'Where we feed in fen and reed, they'll feed both beef and bacon.' It ends with a request for violence: 'We can all agree to drive them out by battle.'[3]

Quite an undertaking for a fish. But it was urgent for the species involved. Mark Cocker persuasively identifies the 'powte' as a burbot; a curious fish, the only freshwater member of the *Gadiformes*, the order of fish that includes cod, haddock and pollock. It looks like a cod crossed with an eel, with a delicate bronze patterning, like sunlight rippling through silty water. In Britain the burbot was found only in fenland rivers before its extirpation in the late twentieth century, having formerly been 'an abundant fenland resident of high commercial value'.[4] For Cocker it makes the ballad 'a piece of environmental art *avant la lettre*'.[*]

---

[*] A posh way of saying something existed before the concept by which we now refer to it.

Drainage spread throughout the fenland region, a slow-burning contagion of destruction, fuelled by our greed for good growing soil until the end of the nineteenth century.* Whittlesey Mere was drained to its inky-black, peaty extinction in 1851. And then the fens were no longer wet and wild. They became this linear, engineered landscape, which has to be maintained through drains and pumps and sluices.

As we've seen, keeping peat wet is essential for keeping it healthy. If it dries out, then the drained fens are a demonstration of what happens next. As peat dries it shrinks; silty riverbeds are now raised up as the land has sunk, a deeply unnatural state of being for water. The minor roads here that have buckled and warped have done so because the peat underneath has dried out over the centuries. The OS map's background here is plain white, devoid of contours. Where they do appear, they are measured in metres rather than the standard 5- or 10-metre intervals. The land also releases its locked-up carbon to the air; the preserved plant matter turns to dust and drifts away, something I have seen once on the wetland edge of Hungary's Hortobágy, where driving through a fen blow was like driving into dusk in the middle of the day. Draining the fens may have released prime farmland but it is easily exhausted, literally vanishing before our eyes for the pursuit of carrots.

Holme Fen, which bordered Whittlesey Mere, is now the lowest place in Britain at 2.5 metres below sea level. An iron post was driven into the soil there after Whittlesey Mere was drained, its top at the ground's level, its base in the clay below the peat. Now four metres of that post can be seen; more post stands above the ground than below. 'Strictly speaking, they are never reclaimed, only being reclaimed,' says Mr Crick in *Waterland*. Whittlesey Mere had to be

---

* Although flood-relief channels were still being dug much later, such as the Maxey Cut in South Lincolnshire from 1950s.

drained again in 1852 after a burst riverbank reflooded the lakebed, water feeling its way to the lowest point in the land, as it must. But the damage had been done.

Tom Williamson, in his history of English wildlife, makes the point that of all the environmental changes that have made the English landscape what it is, 'it is probably the draining of the wetlands which caused the greatest amount of damage'. It left the bittern with nowhere to go. It imperilled the marsh harrier and the bearded tit, a tremendously charismatic bird, mouse-bodied with a long tail that looks implausible for something that lives deep in the densest reeds. Breeding waders began their declines to the point of being on avian life-support. Swallowtail butterflies and fen orchid became marooned in the Norfolk Broads. Some damage goes ignored. Malaria used to stalk these fens. The word comes from 'bad air' – the mist that forms over wetlands was seen as the cause. For Williamson, the loss of the malaria-bearing *Anopheles* mosquito from the newly dry country 'represented, in a strict sense, a reduction in the nation's biodiversity. The fact that it is never discussed as such is a reminder that very few people really believe that *all* of England's native wildlife should be protected.'[5]

Of course, not everyone saw it that way. The fenland poet Edward Storey was pleased by the drainage, writing in 1985 of his pride in the soil and the sea of wheat and barley and potatoes, before doubling back on himself in thought. 'The present has meaning only when seen against the failures or successes of the past,' he writes, 'and an appreciation of any landscape is greatly enriched by an awareness of what has gone before.'[6] The fens are the ultimate landscape litmus test: in one of Vermuyden's cuts and the square green fields do you see creation or destruction?

Variations of that test can be found across the world. We have a good record of American drainage from colonial mapping. When the first European settlers arrived, it is thought that there were

89 million hectares of wetlands across what became the lower 48 of the United States of America.[7] As the colonists spread west, wetlands were reclaimed and drained, the water wrestled elsewhere, the ground dried for agriculture. Or, as Aldo Leopold, the great writer of American nature, put it: 'Progress cannot abide that farmland and marshland, wild and tame, exist in mutual toleration and harmony.'[8] By 2009 that number had halved: an area roughly equivalent to the size of California remained as wetland. Between 2009 and 2019, when we knew full well what we were losing with each lost wetland, another 271,000 hectares disappeared.[9] A policy of 'no net loss' is clearly not working; the then secretary of state for the department of the interior, Deb Haaland, says that the reasons for this are 'complex' without elaborating.[10]

The reasons weren't always complicated. The Great Dismal Swamp of Virginia and North Carolina, which now covers one tenth of the area it once did, was drained for logging and agriculture. Eight million hectares of the Florida Everglades gone for the same reason. The language – 'drain the swamp' – encodes the wrong idea: that wetlands are places of decay and rot and waste. Other wetlands were not spared either: the ephemeral Rainwater Basin of Nebraska has been 90 per cent drained. The dusky seaside sparrow of the Florida saltmarsh went extinct, in part because of drainage to reduce mosquitoes around Nasa's Cape Canaveral base; the last one died in 1987 in Disney World Florida, extinction as a perverse public spectacle. All of this was driven by the same impetus behind the draining of the fens, the desire to subjugate water, controlling land, forgetting that what was there was there for a reason. Leopold, again, was wise on this: 'Some day my marsh, dyked and pumped, will lie forgotten under the wheat, just as today and yesterday will lie forgotten under the years.'[11]

Also forgotten under the years: the Australian cities of Perth and Adelaide subsumed swamps as they grew. Irrigation in the

Murray–Darling river basin is drying out the wetlands, not helped by the increasing extremes of the climate. Though it is not all bleak – 1,000 hectares of the Walker Swamp is being restored after drainage. Sometimes the water is needed elsewhere, to grow necessary food from a hard land, but irrigation threatens to dry out Mali's Inner Niger Delta. Sometimes it's a show of strength: Benito Mussolini ordered the draining of the Pontine Marshes in central Italy and planted a city, Latina, in the middle of it, because mastery of wetlands by brute force is an approach favoured by dictators. Saddam Hussein drained 90 per cent of the Mesopotamian marshes – home to the Shi'a Marsh Arabs – in response to the Iraqi uprising in 1991. In a little over a decade, the NGO Human Rights Watch estimated 210,000 Marsh Arabs were 'arrested, "disappeared" or executed'.[12] It seems unlikely that the Mesopotamian marshes will come back. That land has now become desert and, in a time of climate change, will stay that way.

The story of the fens, though, keeps hope alive. Vermuyden was not the end of them. They still have a way back.

I wanted to be at the fen this November day because Suffolk is wetter than I can ever remember it. The fields by the roads are temporary shallow lakes, shining. The ditches are perilously full of cloudy water, the grundles* grumble with the flow forcing its way through the greenery growing within. Storm Babet had dumped a month of rain over the region in just three days, straight to hard, dry ground. Storm Ciaran added more rain just over a week later. My parents' garage flooded for the first time in a decade after the rains couldn't drain quickly enough. The village where I went to school is

* A Suffolk word that a friend who lives in Grundle Cottage tells me means a weather-borne drainage ditch.

still blocked from the main road by a stubbornly persisting flood. It was as though the earth had woken up to water again. A rekindled desire to undo centuries of drainage and drying. As if Suffolk wanted to rest under a duvet of water and slip into a marshy dream.

The small geographies of my corner of the county – the high Suffolk 'plateau', chalk under the clay; Breckland, sandy and pines; and fenland, peat and reed – all showed signs of the heavy rainfall. But the one with the least amount of still standing water was the wettest: the fen.

From the riverbank the Little Ouse was not so little. Normally it gently flows west from its source at Thelnetham Fen, a low-energy river meandering through its allotted course. Here it has big green banks with a large stretch of grassland between its course and the river walls that I have never seen it need. Beyond the north bank of the river lies a band of open water. In twenty years of visiting, it is the highest I've seen, the water flowing right to the bank's edge, the grassy hinterland soggy and pooled, and stalked by little and great white egrets. A family of whooper swans – ivory adults and dusty cygnets – are bugling and drifting on the open water of the washland. By my feet the body of a drowned mole, swollen grey flesh and slicked black fur, its jaws turned up to the sky, frozen open as if screaming. Its perfect white teeth like tiny chips of bone. But it and a swollen shrew a few metres down the path are the only visible victims, the only thing resembling damage. The reeds are ruffled in part like the fur of a disgruntled cat but there is a different feeling here to the surrounding landscapes: not of a place that was untouched by the storm but of one that is capable of handling vast flows and influxes of water. Reedbeds are like storage tanks. They can buffer big fluctuations in water level and clean the water as it passes through. The river is the drain for the excess rain.

To get to this state took time. Lakenheath Fen began again in 1995, prompted by the loss of the bittern from the British countryside. A squat, creeping heron, libelled by the Bible as a bird of ill omen, bitterns do all they can not to break cover from the reedbeds; their love for reedbeds is so profound that they even look like one, feathered in *Phragmites*. Formerly common across Britain, habitat loss made them rare – and rarity made them desirable. Where reeds remained, bitterns were suddenly a target for shooting, trapping, egg-collecting, skin-stuffing, the culinarily curious. By 1886, records of them breeding had gone.

And then something unexpected happened. Bitterns came back, breeding again in the Norfolk Broads by 1911, although they were never common. World War Two helped. Fears of an enemy invasion meant our coastal hinterlands were flooded again. Reeds colonised these areas. With a nation distracted for six years and habitat unintentionally created, the bitterns gained some momentum, reaching a peak of around eighty booming males in 1954 (they are a species so secretive, the only reliable method of counting them is listening for the males' springtime 'booming' calls, which oversells their faint, hollow-sounding hoot somewhat). Reedbeds are a transitional habitat though. They need care to be sustained, otherwise they dry out and become colonised by trees. They need management like a meadow; the reeds cut, the water level toyed with to rejuvenate the bed and to stop the accumulation of dead reed vegetation choking itself out, lifting itself above the water and letting willow gain a roothold. Reedbed conservation is a process of standing in the way of nature, of holding back the inevitable. Without that effort, the habitat declined and the bitterns gradually disappeared once more. By 1997 there were only eleven calling male bitterns in seven places.[13]

Two years before they reached that low, however, the RSPB had started talking to estate agents about carrot fields. The aim was to create a wetland out of peaty fields: a place that should have been wet anyway and could hold a reedbed inland, away from the coastal marshes where all of Britain's bitterns had retreated to, where they were – are – at risk of rising sea levels and storm surges in winter. By 1997 the RSPB had acquired nearly 300 hectares of land by the Little Ouse. It was time to make a fen.

What they did with the land is what we would now call rewilding. But that term was still fifteen years away from popular use then. It began, in the wild way of things, with planning permission. They couldn't damage the soil. Due to the amount of water the reedbed would be storing above the ground level, they had to obey the Reservoirs Act of 1975 and build an extra 4-kilometre-long embankment of compacted sand and earth between the reserve and the neighbouring farmland, for safety in case the internal embankments failed. Further still, they had not to attract too many starlings or gulls, whose roosts could bring down a jet from Lakenheath's military airbase.

There was also the problem of how to get the water to the site in the first place, as this is the driest part of England. And then how to keep it there without flooding the neighbouring farms, which would immediately pump the precious water away into the drainage ditches that keep the fens dry. Water would be abstracted from the river in winter and kept in reedbeds closer to the river (where it wouldn't flow immediately towards the drier arable ground). A series of pumps was installed to recirculate water on the reserve that flowed towards the farmland.* And it was only then that the diggers could come in and begin to excavate the intricate patterns of channels and areas of open water; only then that the staff could begin to hand plant 300,000 reeds as cuttings and seedlings. Only then did that hard-won open water begin to return to a sea of swaying golden reeds.

---

* For more on quite how complicated this is, see Norman Sills and Graham Hirons, 'From Carrots to Cranes', in *British Wildlife*, vol. 22, no. 6 (August 2011), pp. 381–90.

Fen plants colonised quickly, growing from a seedbank in the soil that had waited 150 years for suitable conditions. Bearded tits returned to breed in 2004. Bitterns took longer, even when the reedbeds had been sculpted to their specifications and stocked with rudd, a small silver-coloured fish, to sustain them; they didn't start breeding there until 2009, as if waiting for everything to bed in first. Even common crane bred on the reserve before then, in 2006. They were the first fenland pairs since the eighteenth century of this most perfect wild wetland bird: lanky and shy and out of scale with the English landscape. Fourteen pairs of crane were breeding in the fens by 2021, a significant percentage of Britain's fragile population. The cost for making Lakenheath fen again was £3,600 per hectare.[14] If they had rewilded the entire fens in this way, the bill would have been £1.4 billion.

Returning water to the landscape begins the slow, millimetre-per-year process of growing peat, the land storing carbon again. Reeds help too. Studies from Denmark[15] and Australia[16] both found that beds of *Phragmites* are effective at storing and retaining carbon dioxide if they were maintained – kept wet and without encroachment from trees – and worked across temperate and semi-arid climates. It should be acknowledged that wetlands give off methane, a greenhouse gas formed by microbes in wetland sediments (perhaps the real 'bad air' of a marsh), but that over time they absorb more greenhouse gasses than they emit.

Wetlands are a buffer. The drained fens used to flood frequently: riverbanks would break and gush water over the land that had shrunk below the surface of the river; pumps would fail and water would back up. In one picture from Lakenheath in the 1930s, two men wade through water that reaches just below their knees, while a third man attends to a truck that is stuck in the water and leaning perilously; the land around them shining grey with a vast amount of water that could be the sea were it not for the hedgerow running through it. The solution at the time was to build more drains. Funnel more water off the land. This November day after Babet and Ciaran shows that by making space for water, the landscape can hold it. It can slow water down and, instead of a sudden spike in river levels, the excess flows back into the river more gently, reducing flash flooding. Wetlands moderate water; drained lands exacerbate the deluge.

Bitterns were not the only species to flourish in the restored habitat. When all of this, from here to the horizon, was fen, the marsh harrier was a common bird. When all of this became carrot fields and poplar plantations, it was not. Their fortunes can be mapped on the same timescale as the bittern: extinct as a breeding bird here in 1899 and recolonised in 1927, only for the population to slip again. Except marsh harrier slid harder and faster than bittern did. By 1971, the reedbeds of the RSPB's Minsmere held the only breeding male in the country.* That reedbed, on the Suffolk coast, was a last refuge from the organochlorine chemicals that were poisoning the world at that point.

From that most precarious position they rebounded. There are now 590 pairs in the country: their range has increased by 884 per cent.[17] That figure is good news of such proportions that we can forget, as another glides just above the bending reed-heads of the fen, that they were once our rarest birds. Bearded tit came close to that as well in 1947, when only four pairs bred: in 2022 there were 847 pairs.

The resurgence in bitterns, bearded tits and marsh harriers (and the gradual increase in cranes) says much about our wetlands now compared to where they were. They are safe havens again: a clean environment free from disturbance, where the wild and the shy can

* Marsh harriers are polygamous and the number of females that the male was breeding with at that time is unknown.

survive. In a habitat that actively repels the human (I have walked through a reedbed so you don't have to and it was deeply unpleasant – dense, sharp-edged, claustrophobic and definitely not made for someone of my dimensions) we need to pay attention to what the birds can tell us about the wetland. Bitterns tell us about the fish, who tell us about the water and the plants and tiny insects that make up their diet, that in turn feed the bittern. Marsh harriers tell us about the small birds and mammals that make up the reedbed and its margins, a different set of small parts in the wetland that build up the whole. Cranes tell us that the scale of our ambition is good – that the landscape is returning to a better, wetter, wilder state than before; the Norwegian novelist Tarjei Vesaas writes in an almost hallucinatory passage of his novel, *The Hills Reply*, of their presence as giving marshes 'a new content, a hidden magnitude'.[18]

This is the land as it once was; the air filled with the birds that were present here, that we used to live in sync with. It can happen again. Mark Cocker critiques the term 'rewilding' for its 'backwards looking and future-averse tone'.[19] But here this makes sense to me. Sometimes the future looks like the past; looking backwards isn't always bad.

Now the restored fens spread hope amid the carrot fields.

After Lakenheath came Bourne North Fen, Ouse Fen and the Great Fen. Bourne North Fen is a 50-hectare project run by the Lincolnshire Wildlife Trust, returning farmland to fen. Lincolnshire, the northwest edge of the fens, particularly suffered from drainage: this small project will enlarge their share of fen by 30 per cent.[20] Ouse Fen is an expanding RSPB reserve with designs on having the largest reedbed in England. At 460 hectares it will be more than half the site's eventual 700 hectares, all growing out of a quarry as Brice Aggregates finishes incrementally digging out the buried sand and gravel that the Ice Age laid down. But the biggest of all is the Great Fen: a landscape-scale project designed to connect two fragments of undrained fen, Holme Fen near Peterborough and Woodwalton Fen near Huntingdon, with a planned area of 3,000 hectares. The idea was first hatched in 2001. Its first land purchase was in 2002 and in 2023 Holme Fen was connected with Woodwalton Fen in what will become, when the restoration works take hold, one broad band of fen for the first time since Vermuyden.

The hope that these places engender is one way forward for the future of the fens. Because continuing with the rest of the landscape in its current state seems untenable to me, as the peat shrinks to oblivion and buckles the foundations of the roads, railways and houses; as the land belches carbon dioxide and the climate becomes more extreme; as the rising sea level reaches up the tidal Ouse and the rivers perch above the lowest point of the land, with disastrous flooding prevented only by the integrity of their banks and wise choices with sluices.

However, rewilding it all is not straightforward. To do so would throw its inhabitants, many of whom earn a living behind the wheel of a tractor or within the furrows of the fields, into unemployment. The lost farmland, which provides 35 per cent of England's vegetables, would need to be replaced elsewhere, more intensively on less efficient land. Unemployment in a region that teeters on depression and poverty anyway is a terrible fate. The irony is that to return all this land to fen would yield the same result as Vermuyden's works: severing the people from their place.

Nobody seems to have a straight answer. It is a topic for a big, awkward discussion, but all I have read is lots of glossy documents offering lots of sunny thinking and aims, committees, steering groups and stakeholder consultations: 'Synergy is sought after in the stochastic future; nevertheless the area will be levelled up and upskilled.' The future of the fens seems to offer a huge opportunity for corporate obfuscation and evasion of responsibility, but little in the way of real action.

It seems to me, in the absence of tangible plans, that easing up on the fenland would be a good idea: reducing the intensity of the farming, undoing some of the draining. Because we need to make space for water. The unaffordable cost for rewilding all of the fens means we have to live with the fens as they are, damaged by the mistakes of past drainage. But we might try to mitigate them while we can, while we wait for the committees to catch up to reality.

There are some places where the past can be seen. Where the land doesn't need restoration and the fen is an example of what everywhere else could be: a refuge for wildlife; safety in the soft embrace of sedge and water and reed.

Dad and I are early for the fen's great show: the roost of birds as the winter sun sets. Fortunately Wicken Fen is the perfect place to kill time. Fieldfares hack overhead from hedge to hedge. Drizzle drifts through a flat grey sky, not exactly as the forecast suggested. A marsh harrier glides through the cold afternoon air in its lazy V. Two boys carrying fishing rods get to the edge of Wicken Lode – an old man-made waterway running through the fens – before exclaiming, 'Shit, never seen it so high.' Me neither. What remains of the path is the good black soil of the fens, sodden from Storm Gerrit and churned by feet to a sticky, slippery slick. We slide over it. Mud and water mean life. The breeze brings the whistles of wigeon and teal from the open water we can't see. The gurgle of a moorhen and the squeal of a water rail from the reeds. A reed bunting clings to the crown of the hedge. In the lode, the pike – traditional winter fishing – stay still, not opening their snouts to seize the boys' fishing hooks with their fierce teeth. These fens are rich places. Predators can bide their time.

My dad can't remember when he first brought me to Wicken. But, he says, nodding at the months-old-baby grumbling from a nearby birder's papoose, I must have been a similar age. So if Lakenheath Fen is buried within my head then Wicken Fen goes deeper still; it is under the skin, the air familiar to my lungs, the memorable soft feel of it under my boots since I began to walk, the chill of it in my bones. It is a place that in this particular set of circumstances – cold winter afternoons – is as well known to me as the family home.

Wicken Fen is a series of fens with a chequered past. Sedge Fen, the main wetland, was never drained: it is a wet island in the fallen fenscape, and in 1899 became the first nature reserve in the care of the National Trust. It was, at the time, rife with relicts: species that had been lost elsewhere through drainage, such as the swallowtail, a palm-sized butterfly in yellow and black; the fen violet and fen orchid, two small flowers with very specific needs; and the reed leopard moth, a large brown moth that in its larval stage spends two years inside the stem of a reed, below the water line. Now only the moth and the violet can still be found here. The fen's proximity to Cambridge University meant it was well studied, in turn making it a magnet for amateur naturalists of the period.

Adventurers Fen, on the other side of Wicken Lode, was drained and worked for its peat. When this became uneconomical it was abandoned. Water, as it will, found its way into the workings, sparking a reedbed into life until World War Two when it was requisitioned, drained again, and the soil borrowed for growing food for the war effort. After the war it was returned to the trust, the peat allowed once more to settle under water and reed. Neighbouring Burwell Fen beyond has been bought, also with the intention of bringing it back from agricultural land to wetland.

The need for it is clear if you take the road to Wicken from Soham. It is a land ripe for the mind's rewilding: 16 kilometres between the horizons, from Ely Cathedral, dominating the northern horizon like a great mountain peak, to the gentle swelling of chalk at the end of Suffolk. All that lies in between seems to be grey-green agricultural fields without end, skeletal hedgerows and a whole lot of space; as if the landscape has sublimated straight from solid to air. But then you turn off the road and find your way to the edge of the fen and see how it could all look; ten thousand years of environmental history condensed to 130 hectares, the missing wet step in the landscape.

An hour to go. A bog oak found in 2016 lies next to the visitor centre, pickled by a few thousand years in the sodden soil until the wood is dyed black as the peat; fissured like a gargantuan chunk of charcoal. It is decaying as it dries out (even after all this rain), shocked by the sudden flood of oxygen, a substance it hadn't felt since it breathed Bronze Age air, photosynthesised Bronze Age light. It marks the threshold to an extraordinary place.

Beyond the visitor centre, the fen. A white-sailed black windmill sits stationary in one corner. Beyond it, the bare birch branches, the tallest in the woodland, blur to a kind of single dark grey mass, a smoke that sits, moving in the breeze without ever going anywhere. Above, clouds scud below other clouds, the early afternoon rain clearing, blue creeping back into the corners of the vast sky, though the bulk is still billowing the shade of grey that promises snow. You always notice the sky in the fens, perhaps because the land is so low that it is like standing in the nave of a cathedral, looking up at the high, huge ceiling, reverently admiring the architecture of the light. Under all that sky: sedge. Sedge in December the colour of a barn owl's wing; huge beds of it like a shorter, less sturdy version of the *Phragmites* that line the channels of this fen.

Saw sedge, also known as fen sedge, has leaves as thin as paper but as rough as a cat's tongue; they branch from the stem on a regular basis and remain strong even when dead and dried. It is ideal for thatching. Sedge defines the fen deeper than just the name. Thatching is what kept the fen fen and the National Trust claim it's the reason why the local community resisted the draining of Sedge Fen.* The sedge is cut on a rotation, after a patch has grown for four years, so it can be harvested without weakening its grip on the fen. This is what ecologists refer to as a deflected succession: where the plant community has reached a stable state that it would not have if it wasn't for the continuous intervention 'deflecting' the progress to a different habitat.† Without the harvesting of the sedge, the fen would have long disappeared beneath a carr of birch, willow and alder buckthorn (which has been harvested here too, for its charcoal is ideal for use in gunpowder and fuses).

The boardwalk we are walking runs parallel to the lode, lifting us over the soft ground where water pools below the boards. It takes us to a corner where the sails of a turbine spin fast in the breeze. A derrick lifts and falls with a metallic squeak. It is a pump but it is pumping against the flow of history, taking water from the lode and returning it to the fen. It is a form of life support. Without it, Sedge Fen has been so isolated for so long from its main source of water, the big winter floods that would have covered the fens, that it has been gradually drying out, the peat oxidising, leaking carbon. Even ancient fens need rewetting, even in wet winters like this one. Looking after a fen is continual work; a continual process of standing in the way of progress, where water and plants and time are suspended.

The clouds part, the day brightening just as it ends; the banks of grey pushed to the sides as if preparing the sky to be a stage. Those

---

* Plausible but I haven't been able to verify this.

† The theory of a deflected succession was first developed by Harry Godwin, a Cambridge botanist in 1929, based on his puzzling out of why Wicken Fen looks the way it does; it is a little piece of this fen immortalised in ecological theory.

that remain change colour. Grey to white to butter to peach. It begins. Lines of gulls seeking open water elsewhere wobble through the sky; then a wave of lapwings, wings as wide as human hands, in silhouette against the light. Greylag geese honk. The first small flocks of starling break cover, a hundred flickering wings driving up into the sky. A cormorant draws the black arrow of itself towards the setting sun.

The blue darkens now by increments: sky, cerulean, cobalt. The sedge fades from view; the taller *Phragmites* heads just lifting above the dark horizon trees like a stubble, a fenland four-o'clock shadow. The starling flock thickens into the thousands. They swoop back, a tight bunch that as it plummets loosens into a flow of birds like an unravelling ball of wool, the thread pulled fenwards. A marsh harrier drifts, almost aimlessly.

And then: a pair of black wings flicks up against the dying sky. I have seen enough now to know it on instinct; some almost imperceptible distinction of the shape as it spins on its axis in the sky and turns away; the body built lighter; the wings thinner, angled back at the carpal a little more, wings used to a speed and purpose of flight that marsh harriers aren't: hen harrier. I keep tracking as it slips from silhouette to skim below the tree line, its brown feathers scarcely visible in the dark, its ring of white at the top of the tail a flash in the dark. It splits in two. One banks right, back across the fen, while another white flash drifts left along the treeline. Frigid fingertips forgotten.

After a summer's solitary breeding in the uplands, hen harriers winter in the fens: they live peaty lives, finding their niche in both seasons in these open spaces, hunting small birds and smaller mammals. To find one in the daytime is a stroke of luck. It will disappear, effortlessly finding its way to the horizon. Sleep is the problem. They want depressions in long vegetation; where they find one they can be a bit more social than earlier in the year. It is the abiding thrill of the fen: a place of food and safety for the night, for this, Britain's most threatened, most persecuted bird of prey, the peatland predator.

A third and a fourth: over the sedge two silver-grey and white males materialise, joining the two ringtails.* The males have the ghostly plumage but against the dark surroundings of the fen, they show up; stark and bright, as if they absorb and reflect the last of the light; whereas the larger, dominant mottled brown females are the ones that move like shadows. They all drift out of sight.

Light almost gone. Half an hour past sunset and day is remembered only in a faint puddle of lemon and apricot where the sun disappeared. The starlings swirl once more, a great wave flying low over our heads before disappearing into darkness. Scanning the fen for harriers I hear something else. That unoiled yawning of metal on metal coming from somewhere other than the pump. I lower my binoculars. A distant black speck in the sky, disappearing west. I raise them. Two birds. Big birds. Necks outstretched, legs trailing, wings flapping slow but with purpose. Cranes. And within seconds they have sailed through the colour, through the visible sky, beyond the western horizon. Even fleeting, even distant, even disappearing; their presence is enough. We respond with silence. Wild reverence.

Dad and I nod at each other. Things don't get better than Wicken in rhapsody. But it's time to follow them. We drift back to the car in the last of the light. A Cetti's warbler erupts in song from the undergrowth, reverberating around the fen now stripped of other noises. The pools of water beside the boardwalk are now the deep blue of dusk; everything else black but for the light spilling through the visitor centre window. We take one last look back. One of the male hen harriers swings slowly around, now also a shadow in

---

* A catch-all term for the similarly plumaged females and young birds.

the gloom, before stalling in the air, and plunging feet first into the black sedge of the fen, its safe roost for the night sorted.

And to think, but for the sedge this too could have been just another carrot field, another footnote in the fenland obituary, those hen harriers merely ghosts. Instead, at Wicken Fen, the past persists, offering a picture of a potential future for the rest of the fens: what they could be and how they could look if allowed to be wet again; if the long-buried seeds in the soil were given the conditions to sprout, the peat allowed to absorb carbon instead of drifting away in the breeze, the reeds absorbing and cleaning fluctuations of storm-water. A restored fenland, sparkling with life.

The reeds are dew-slicked. Each one wrapped with a halo of golden light; the bulrush standing stiff, the Phragmites drooping with the weight of water. The still air stirs only for the ping ping calls of a bearded tit, notifying itself to the morning.

At the reed face, there is nothing. And then a sudden something, a bird appearing where it looks like no bird should be. One foot on each reed stem, the rotund ball of a bearded tit's body suspended in space, as if walking the tightrope of its own legs. Shadowy blue-grey, two long black markings descend from each side of his small eyes and stubby, stuck-on beak; his beard merely a moustache. He flies to a tray of grit. Swallows a few small stones. After a summer of insect-eating, he needs the grit to help grind away the tough casings of reed seeds, his winter diet. Turning sideways, his body changes. Apricot, a colour warmer than reeds at dawn, replaces the grey of his head. Wings that look too thin are stuck on his side, and a tail, ridiculous, like a fragment of reed stiffly sticking out of his backside.

A female flits in next to him, stripped of his frippery: reed brown and black-shadowed wings. Then they're off – slipping back into the fringe of reeds, finding space where it doesn't seem to be, their long tails somehow flickering through imperceptible gaps. They slip along the edge before flying across – ping ping – disappearing into the depths of the reedbed, not to be seen again.

Strange things make sense at dawn. Before the cold light of day kicks in, it seems normal they should live here, in the depths of the Phragmites, an environment with its own peculiar logic, as strange as the deep sea. A bird with a moustache that pings like a phone; a bird that has to eat stones; a bird with a long tail that lives in a place without the space to swing it. I don't see another all day; as if they were dawn's apparition, as invisible as bitterns, as invisible as the spaces between reeds.

8

Between Land and Water

Returned to the river, the droplet grows saltier by each metre. It tangles with the push and pull of the tides. Its final obstacle. At the last, it is swept over the edge of the land by the highest tide, pushed into pools and creeks, over stunted grass and strange plants; the tide turns and the droplet ebbs away. To the sea; to the end. Ready to begin again.

Nowhere I have been resists being loved like saltmarshes do.

I am standing, sweltering, on a hummock in Wigtown Bay, salted grass under the soles of our wellingtons. We are wearing our coats because there is nowhere else to put them, while the mudflat in front of us waltzes with heat haze. The top of my head is turning red. When we arrived this morning the sky was flat grey. Cairnsmore of Fleet, the biggest mountain in this coastal corner of Galloway, standing across the bay, dominating the horizon darkly, the lilac and lemon stars of sea aster by our feet. A sudden shock of rain swept sideways across the bay, stinging. August in Galloway can be cold and wet or hot and dry. Or, today, both. We were a little way across the marsh when the clouds burned off. Now the mud is crackling in the summer sun, releasing sulphurous gases from decaying matter that can't be seen, only heard and smelled. This is the last land that the rainfall will see before sweeping out to sea, the cycle complete.

There are bigger saltmarshes. There are wilder saltmarshes. But rarely are they as interesting to me as this saltmarsh, wedged between the tongues of two rivers, on the shore of a firth; harboured and historied. Wigtown Bay also happens to be Scotland's second largest saltmarsh, created by accident and almost visited by thousands every day for a fortnight, as September turns into October. But while Wigtown's book festival brings people to the town, they rarely make it to the hide down the hill that overlooks the buffer of saltmarsh between the town and the sea. The literary rarely meets the littoral.

If I was to be asked why, I'd say that in comparison to the other waterlands, saltmarsh is the least approachable, the hardest at first glance to understand instinctively. If you come here, you could find yourself looking at a green field or a grey sea or any of the states in between. It could be barren or bustling with life; firm under foot or fluid and flowing; washed with freshwater or flooded by the sea. Saltmarsh is an either/or sort of place but sometimes it is both: the final word in the permeable boundary between water and land.

It is the transient twin of freshwater marshes. With fen, the freshwater is consistent, which allows reeds and sedge to grow to a great density and height. The area of waterland is clearly defined. But here, where the water is spiked with salt and fluctuates wildly, the vegetation is stunted and variable. Paradoxically this waterland is well defined by distance and tide but ill defined by foot and eye. In its mixture of water and mud, there are times when it can be alive with birds and invertebrates, and times when you'll never see a thing, with no clear reason why.

Saltmarsh is a sort of land that grows out of the sea. It takes a plant to get things going, and from there things accrete. A seed, caught in the shifting mud, exposed to air at low tide and covered by water at high tide, stirring. It needs to be a halophyte – a salt-tolerant plant – or it will not survive its twice daily soaking by the sea. Something like the samphire* that you find near the supermarket fish counter will do the trick. It grows in both directions, fleshy and green, bursting through the surface mud while its lacework of roots burrows below, creating a sort of botanical net both sides of the surface.

The waves and tides keep rolling into the bay, keep ushering in sediment that gets trapped by the new plant life. Formerly transient, the sediment now builds up. As it builds, more plants grow – land in the act of becoming land. Land caught in between states. Later the sea will come and reclaim it. Tomorrow everything might have changed, the water working away, eroding in parts, making anew in others. Out of the muddy wet chaos, a saltmarsh is born.

'Most people don't like being out in open landscapes,' Elizabeth says. 'But I love it here.'

I do too.

Elizabeth Tindal knows these marshes well. A local, she was once the council ranger for Wigtown Bay, Britain's largest local nature reserve. After redundancy, she started a business doing all the fun bits from her previous job, which is how we both came to be standing here, surrounded by stinking, shimmering mud, on the last hummock of grass. Elizabeth bounces with energy. The Tiggerish glee of somebody paid to have fun in a place they love.

I take in the view. We are on the western shore of the bay. East, through the haze, a small expanse of sea can be seen running where the channels of two rivers, the Cree and the Bladnoch, join. The view north and south is over a plain of green saltmarsh, speckled with

* Supermarkets never spell it out but this is marsh samphire, *Salicornia europaea*. The other samphire – rock samphire – is unrelated and grows on rocks. I haven't eaten it but it is still edible with an apparently funkier flavour.

sheep. Perspective erases the middle ground to the north, where miles of hinterland separate the marsh from the highest hills of Galloway. To the south, marsh seems to hug the flank of the peninsula that sweeps out to the open sea of the Solway Firth. Back over our shoulders, the spires and roofs of Wigtown rise above the wooded slopes at the back of the marsh, as if the literary town is unwilling to get its feet dirty in the saltmarsh below.

But this is the wrong direction to be looking. The real view is down below, on the scale of the grey-toned slick of mud that shines with water. Closer still. There are dimples, pockmarks and tiny lines written in the mud, each one alive with meaning, suggesting the strange habits of invertebrates, alien lives lurking. The marks of this landscape are illegible to me though, words that I don't yet recognise. I am illiterate in this language. The mud might as well be a blank page.

And this is an unpleasant feeling. But Elizabeth is handing me a yellow fork, the tines the size of my hands – the paraphernalia of a proper saltmarsh naturalist – and telling me to dig. And I know whatever I pull up will be utterly baffling.

The influxes of saltwater require special kinds of species to cope with the conditions. But because the seawater doesn't reach across all of the marsh on every tide, it's a habitat that changes with distance, by gradients as you progress further inland, where the sea has less influence. It can be laid out on a graph, the *x* axis mapping the distance from the sea. At the seaward edge, it is pioneer marsh, home only to the specialist plants such as samphire and cordgrass that are the first to grow, creating land from the sea, allowing the strange invertebrates and the birds that feed on them to gain a foothold in their wake. The further back from the sea, the less salty the saltmarsh becomes, and a greater diversity of species will be found, able to tolerate the gentler conditions. The more fixed the mud, the more conventional it will be, and the more familiar, general species you will find; there will be freshwater insects in the pools, the grass grows thicker and longer. Although convention is relative in a place as extreme as a saltmarsh, where living is always stressed to some degrees by the salt and extreme changes in water levels.

This is a generalisation of course. There is no singular, definitive saltmarsh. Some, the ultimate exception, are found inland where old saltworks have flooded. Some saltmarsh plants such as scurvy grass – so called because bloody-gummed and bruised sailors would eat it to ward off the effects of scurvy – grow along gritted motorways, flowering white in spring as auxiliary road markings. Definitions of saltmarsh are a nebulous fudge – any definition more than salt, water, mud and specialist plants tends to hide as much as it holds.

Like saltmarsh, Wigtown Bay as a place is hard to get a handle on. The nature reserve here is 2,854 hectares in size – only a little bit smaller than Heathrow Airport – and it is a lumpen, roughly triangular expanse of mud, with two tadpole-like tails extending from it – the rivers Bladnoch and Cree. These tails make it tricky. There is no one spot from which to view the entire reserve: the curves of the river, the kinks in the land, the diffuse nature of the mud that makes up most of the reserve sees to that. The rivers are the only definition in the shape of the nature reserve: every other border for it feels arbitrary in the blank open space of the bay.

Water comes from four directions into Wigtown Bay. You can draw a line on the map to delineate the protected area, but water abhors a boundary. When it sweeps east up the Solway from the Irish Sea or the Atlantic, the idea of the bay as a single space gets complicated. Rain falling in the Galloway hills to the north or over the border in South Ayrshire will flow into the Cree. The Bladnoch's watershed is rotated through 90 degrees west, draining the top of the Machars peninsula to the west. Rain that falls over the northern Cumbria, the border hills, the southern half of the Lowthers, all flows into the complex cocktail of water that is the Solway Firth.

If the Lowther hills are, as I described earlier, a bagatelle board for water then here is the opposite. No chance involved, there is an inevitability to water here. It all ends up together in the Solway, giving and taking sediment from this saltmarsh.

We drive our forks into the mud. Elizabeth with the certainty of experience, me with the slight hesitancy of the beginner. As we lift them up, the earth belches sulphur, the mud parting in lumps, cracking along the lines of the invertebrate burrows.

'What have we got?'

Elizabeth begins her translation work. The blank page comes to life. I catch *Corophium*. The little dark dimples in the glistening mud are signs of their presence. The teeth of the fork that bit into the mud is now a wriggling mass of mud shrimps, their grey-brown bodies segmented, crawling through the wet mud on squat legs. Woodlouse-like but for their two large antennae held out and at an angle like a flicked V, as if registering their displeasure at my disturbing them. They scatter. Back to the mud. A redshank – brown-grey body and walking on bright red, compass-point legs – would eat the lot, if only I left them alone on the surface here. But I am going nowhere yet.

Elizabeth has done better. Her fork has excavated a radiating pattern of star-like lines in the mud and revealed a clam. That star pattern suggests it could be Baltic tellin, a small clam of northern salt water and apparently a good one to find. This clam is not one of those, but we reach the limits of our knowledge with it, so we are left instead to excavate what we can observe about it: it is the size and almost the shape of the tip of a thumb, off-white in colour, and a few seconds after being returned to the mud, its foot – a small fleshy appendage – appears, which it will use to force itself back to safety, under the surface of the mud. These are just two of the invertebrates that lurk here, usually unseen, requiring a hands-on approach, leafing through the pages of the mud to find. Out there under the cover of the surface, there will be millions more than the handful we find here.

This mud – sulphur-belching, glossy and slick – is interesting mud. Below the wet and shining surface it is solid and dense, black in the shadow, graphite in the sunlight and riven with rusty stains of ore. It is one of the supposed explanations for where the bay's local name, 'the inks', comes from. It stinks of sulphur because it is anoxic. The tide washes the surface with water and oxygen but below that surface layer there is no air, just bacteria. A neat circle: the decaying invertebrates sinking into the soil, sustaining the substrate that their future generations will burrow through before the beak of a curlew slides in and snuffs out a life.

That is not the only interest in the mud. It holds an archive too. Mud generally builds up undisturbed, and as it does so it forms a record of our mistakes and bad decisions. The sediment of Wigtown Bay is gently radioactive. It is in part unavoidable: a baseline from the granite, from the background radiation of being alive. But there are other man-made sources. One is the Caesium-137-laced rain that swept Europe in the aftermath of Chernobyl's great tragedy. Not acid rain but radioactive isotope rain. Another source is over the border, lurking just around the corner of the firth. Sellafield is Cumbria's nuclear power plant. When it discharges particles, they get swept by longshore drift around the headland and into the firth where the current ushers them into the bay's bottleneck of sediment. Americium-241, Strontium-90 and Caesium-137 get laid down into the mud by the same process that forms saltmarsh. SEPA consider the risk of premature death from this to be unlikely – as if this was the only possible concern.*

* The effect of radiation on the natural world is hotly contested: Chernobyl itself is held up as a rewilded success story, yet the booming populations of birds there have smaller than average brains. At Wigtown Bay the effect is unknown.

Being out here beyond solid ground is a privilege granted by the tide. To get here I had to learn a new way of moving. When the solid ground runs out, you have to glide like a surreal skater, the mud as your ice rink. Elizabeth tells me to keep my weight over the toes of my front foot and to push off with my back foot and that – if done successfully – I would take off in a glide. She demonstrates gracefully. I attempt to copy, gracelessly sliding, windmilling arms, panic in my eyes for the first few gliding strides. But it works. We stop. Another tip. The trick to not joining the creatures below the surface is to stay stock still, an exact, heron-like stillness, or gradually you begin, imperceptibly, to sink. I was glad of her guidance.

Back from the mud we meet those pioneer plants that begin to turn the sediment into saltmarsh. The samphire is not much to look at. Just a few crooked green fingers growing out of the mud. It's hard to credit it as being the plant that can begin all of this.

Samphire has the qualities of a succulent, holding freshwater tight in its waxy skin, away from the salt and the sun. It's why this particular halophyte is delicious when foraged, tasting of the sea, fresh and salty. Richard Mabey describes the taste as being 'redolent of iodine and sea breezes' due to their adaptions to the sea. If you try to keep it in fresh water, he says it 'sucks the sap out of the plant'.[1] The taste of samphire is the taste of a life adapted to the littoral: the mud and the salt and the harsh sun that characterise the pioneer stage of a saltmarsh.

During the Wigtown Book Festival of 2019, as NASA's Landsat 8 satellite passed overhead on a day of startling clarity, it took a picture of the complete Solway Firth, from the twin tails of the Esk and Eden rivers in the top right corner to the Isle of Man in the bottom left. The colours in the published image have been manipulated: the Solway sea does not really possess such a mother-of-pearl lustre, the mud is not really such a shade of Irn Bru. But the colours, pleasing though they are, are not really the point. The manipulation highlights the feather-like flow of mud, the tonnes and tonnes of sediment drifting through the firth, from their origins far out of shot. 'Sediment' comes from the Latin *sedere*, meaning 'to sit', a shared root with 'sedentary'; it suggests that sediment is something stationary and solid, dull and unadventurous. But from its orbit in the dark abyss of space, staring down at the mud and sea and fields that seem to shine like jewels, the satellite shows that is not the case.

The same perspective, two-and-a-half centuries earlier, reveals how much it moves. When William Roy was compiling his map, the scale he was using does not reveal much of 'Wigton' as a town, other than that it being much smaller then. What it does show, perhaps because of its origins within the military, is an incredibly detailed picture of the bay, the texture of the land painted onto the paper, rather than outlined in contours. It is of interest to us crucially for what it shows of the sediment and the water of the bay. If you stand where the harbour car park is now, you see the Bladnoch gently snaking into the sea in a northeasterly direction. Roy's map shows us the Bladnoch kinking back on itself, turning northwesterly to run parallel to the town, before gently bending to the east and joining up with the bay. The harbour then was halfway along the marsh, close under the wooded slope at the edge of town. The shifting sediment of the Solway saw the end of that harbour.

The Earl of Galloway was the man who would move a river. In the nineteenth century he built a series of breakwaters that channelled the flow of the Bladnoch in a new direction. It was a technical innovation. The harbour needed to cater for the new steam-packet ships, with a constant, clear channel of water; the old, silty harbour wouldn't do. Ponies pulled the grey stones to the riverbanks where labourers broke them down, shovelling them into place on the thin crust of mud.

On the one hand, it was successful: two hundred years on, the river remains in place, the breakwaters like ribs, the stones yellowed by lichens, the gaps filled in by mud and grasses, keeping the river in the earl's chosen place. On the other hand, sediment is still brought in on the high tide stifling the new harbour too, which as a result sits mostly unused for anything other than its car park. Saltmarsh has formed inside the breakwaters, squeezing the river, as well as outside of them. As the nature conservationist Clive Chatters writes in tediously correct language, 'The intertidal habitats of the Solway are highly dynamic. The merse [saltmarsh] is in a constant process of accretion and erosion; such changes are integral to the life of a saltmarsh.'[2]

The dynamism of nature – the liveliness of water, mud and plants – defeated the earl's economic dreaming. This place is a unique happy accident: a natural process and someone's failure.

We have moved away from the pioneer phase, the slick and the samphire, to the plantscape that lies immediately landward. Elizabeth threads her way through the lacework of saltpans: the creeks and pools that give saltmarsh its amphibious feel. Now grass – actual recognisable grass, nothing too weird about it – forms on a thicker, more stable crust of mud. As we move inland it thickens and lengthens, growing greener than it did at the edge of the marsh.

Elizabeth beckons my attention down again, over the edge of a creek beyond where a clump of scurvy grass grows. She has found a weird grass: a patch of *Spartina anglica* (or common cordgrass). This is a plant with a complicated history. It looked innocuous enough here: a reedy grass, green and flat-leaved, standing a little taller than the grass in the bay and running along the edges of a channel. The leaves are held tight to the stem in sheaves but branch off at sharp angles and the effect is rather charmingly described in a NatureScot technical report as making it 'look like a slim corn-cob at a distance'.[3] It is behaving itself here, as it does across much of its range in Scotland, but this is not the case elsewhere. It was first discovered in the saltmarshes of the Solent, around Southampton, in 1882. Thought to be a species new to science, it was later found to be a hybrid of small cordgrass and smooth cordgrass. It was infertile, as hybrids should be – then suddenly it wasn't. The hybrid had evolved, developing extra chromosomes in a process known as allopolyploidy, and with it, the ability to reproduce.

Hybrid vigour allows it to spread quickly. Human endeavour allows it to spread further. *Spartina* have the capacity to grow in the pioneer zone (though here they are growing a little bit further back than that). Common cordgrass was a miracle plant for being able to create saltmarshes, literally to grow land out of mud and sea at speed, which could then be used for grazing, farming or development (the fate of many saltmarshes in the Solent was to be swallowed by Southampton). In 1963, common cordgrass was introduced to a single site in China to assist in land reclamation. Twenty-one plants survived their first year; twenty-two years later 36,000 hectares of Chinese coastline was covered in it.[4] Subsequently, smooth cordgrass, also introduced, has taken over from common as the dominant invasive species there.

Common cordgrass is just an innocent plant, obeying the demands of its transgressive, hybrid biology. But when it takes root across the tidal waters of the temperate world, it builds up, a rampant spreader to the detriment of everything else, even to the mudflats itself. It has even been too effective for its own good, accumulating so much silt that it starves its roots of oxygen and dies off. In Scotland it is present in large numbers only on the Solway, where it has yet to show any detrimental effects. For how long the plant will remain placid here is unknown.

As the saltmarsh solidifies, the creeks grow deeper, offering more established passages for fingers of tidal water to creep into the marsh. But the land is fickle and can block them off, creating pools that are salty on sunny days when seawater left by the tide evaporates; or after heavy rain, fresh enough for insects such as the lesser water boatman to survive. The land may be fixed firm under our feet this far back, but the influences of the tide and rain, salt and fresh, still exist. That either/or/both of saltmarsh penetrates deep. And apparently the ground might not even be as solid as I think.

'My general advice is don't get too close to one you can't see the bottom of,' Elizabeth says as we make our way to another saltpan.

'What, in case it's six-foot deep?' I laugh.

Apparently so. The tides, again: the action of the water flowing in and out cuts at the soft mud. It scours out the bottom but also the banks, creating overhangs. The matted grass clings to whatever thin skin of soil is left underneath, an ersatz bank that Elizabeth tells me will collapse under the lightest footfall, dumping you in the water. This creek turns out to be safe. I am not pitched head first into an ad-hoc cold swimming lesson.

Elizabeth takes a clear plastic cup, fills it with creek water, tinged yellowish with suspended sediment. We pull out the other important equipment of the saltmarsh naturalist: small nets. We dip them into the water and sweep. I catch nothing. Elizabeth catches a fish, half my little finger in length, and seamlessly slips it into the cup (by now I am used to her doing everything better than me).

I like to be specific but this environment is a reminder of one's limits. We can't identify the fish. It is tan brown in colour, not dissimilar to the colour of the water in the cup, dappled with markings a shade darker. It tapers elegantly to a point along its body from its widest, at the gills, to a translucent tail. Its head is snub-nosed, almost frog-like, bulbous eyes prominent, raised from its forehead. Its pectoral fins are clear, like the tail, and shaped like diamonds.

The limits of our experience deny us the awe that comes with knowledge, of being able to connect it completely to its role in the saltmarsh. In this place, which is deeply ordered and organised into structures, it is a reminder that not everything can be neatly filed.

A ripple of shadows catches my eye. The motion is strange, difficult to discern, but definitely movement. Camouflaged living things.

We swish our nets through the water towards each other. Elizabeth nets seventeen prawns. I get one.

'They're really skilled at escaping,' she says, kindly.

Elizabeth empties her net into a special viewing device. A couple leap out before the magnifying top is placed on it. I quietly return them with my lone prawn to the water.

They are hard to get a clear view of before the magnifying top goes on, their bodies not camouflaged but translucent. They're an uncanny thing, as good as ghosts for being alive but see-through, a fleeting, flitting presence: solid yet disappearing. My camera records them as prawn-shaped blurs between the surface and the substrate, as though my lens was haunted by multiple slight smudges.

Under the magnifying lens they become defined only by their essentials: the darker edges that give them structure and the dark line that runs through the middle of their antennae. Each has a pair of staring eyes, sticking out at the side of their head, spherical on the outside, flat on the inside and appearing as one large black eye divided between each side of the head. They sit above a smudge of dark internal organs, and the thin dark line of the intestinal tract. Although 'sit' is not a word prawns understand. Even in their temporary Perspex prison they behave as they did in the saltpan: boisterous, clambering over each other, a riot in the water. Their downward bent bodies are muscular, and they flick them backwards as an effective escape mechanism.

It quickly becomes a habit, staring into the puddles, attempting to decipher life out of murk, sweeping a net to reveal what we miss. We catch another *Corophium* mud shrimp. It is like a different animal to the ones we found earlier crawling through the mud. In the water it swims with its segmented body curling up and flicking forward, while its two huge antennae are still extended – they look like a swimmer's arms, as if the shrimp might at any moment front crawl forward through the water.

These invertebrates – these alien life forms – are what calories look like in the saltmarsh diet. They sustain a complex network of life: up into birds; across into fish; below, decaying into the sediment and enriching the soil that the marsh grows from.

We carry on walking, heading back to the harbour, up the strata, into a richer underfoot flora, a richer micro-landscape.

The mud that underpins the saltmarshes does not only hold the records of our nuclear mistakes. The way that it builds up forms a carbon sink. The back of a saltmarsh is essentially a layer of grass held over a massive store of locked-up carbon. Unlike dry land – eroded, developed, disturbed – untouched saltmarshes keep putting down new layers of sediment: the process is the same as it is in the bogs and fens. When most plants die, the carbon they have absorbed is released back into the atmosphere. When saltmarsh plants die, they become buried in the sediment, decaying slowly into the soil. The carbon that leaches out is stored in that soil, undisturbed. On average, 6 kilograms of carbon is stored in every square metre of saltmarsh.[5] That's 370,000 tonnes of carbon across Scotland. Wigtown Bay holds 40 million kilos of it across its 676 hectares of saltmarsh. I find this fact hard to digest, but it is the equivalent of growing 2.4 million trees from seed for ten years, which equates to 960 hectares of commercial forestry.

The saltmarsh here has an excellent view of the Galloway hill flanks that have disappeared under pine plantation. Saltmarsh has the same effect on carbon in two-thirds of the space, if only we choose to keep these areas intact, undisturbed.

Mud and saltmarsh, for ever unloved, have always been easier to destroy than to preserve. Too often saltmarshes have, like bog and fen, been reclaimed from the water for agricultural land or for building on. Not only is this habitat destruction, but it allows the carbon inside the saltmarsh to be released. NASA's Landsat satellites enabled its scientists to calculate the rate of global saltmarsh loss over the last twenty years. They totalled up 1,453 square kilometres, releasing 16.3 million tonnes of carbon dioxide: the same emission as 3.5 million cars. Historic saltmarsh reclamation has given us places like Fairbourne, in Wales, a village built in the nineteenth century that climate change is expected to render uninhabitable by the middle of this century, when it will no longer be defendable from rising sea levels. As with the fens, when the chance of a catastrophic flood exists, it is not wise to chance it.

Most European countries with coastal wetlands* have lost between 50 and 80 per cent of them.[6] The Wadden Sea coastline – shared by Denmark, Germany and the Netherlands – is where 20 per cent of European saltmarsh is found. Yet this is merely a 'relic'[7] of what was once there, when it was a flat tangle of creeks and mud and plants; before the era of polders and dikes. Sea-level rise is also expected to drown the Wadden Sea coastline and its marshes. In China, 24,000 hectares of coastal wetland was reclaimed every year between 1950 and 2000, adding up to half of its coastal wetlands being lost.[8] And while there are movements to restore other coastal habitat in China, saltmarsh gets forgotten.

But short-sightedness is a human trait: plans for non-future-proofed places are still being drawn up. Portsmouth City Council

* A slightly baggy term that includes saltmarsh, seagrass beds, oyster beds among others.

is proposing to build 3,500 homes on the saltmarsh peninsula of Tipner West. The council's logic is that to leave the saltmarsh would let the land flood. Homes, famously, never flood. It seems that people think the only way to save land from the sea is to build on it, though this forgets that the sea will always want to reclaim land back for itself, especially at a time when the waters are rising. Portsmouth Council could learn from Fairbourne's example.

Or they could learn from further afield. Two of the great conservation disasters of the twenty-first century took place in 2010. One was the Deepwater Horizon tragedy in the Gulf of Mexico, where the gushing well washed crude oil ashore across a 1,050-kilometre sweep of coastline. Of its many devastating effects, the saltmarsh of Louisiana was particularly hit. Any oil kills the marsh vegetation, weakening the marsh, allowing the waves to wash it away. The vegetation, hardy, can grow back quickly if the soil remains. But a weakened marsh is an opportunity for strong storms to cause maximum damage to the buffer between sea and land.

The other disaster happened in South Korea with the reclamation of the Saemangeum mudflat. The environmental journalist and nature writer Michael McCarthy writes movingly in *The Moth Snowstorm* about a visit there with Nial Moores, the director of Birds Korea, to what they both now saw as a 'deadscape'. The mud is to become either agricultural land or concrete; the birds it has displaced, the Nordmann's greenshank and the spoon-billed sandpiper, are both critically endangered, already on the trajectory towards extinction before the loss of this precious mud. We don't have either of these species here, but we have their close relatives, our own endangered waders of the mudscape: the redshank, the dunlin, the curlew. McCarthy poses the question, 'Who writes elegies for estuaries?'[9]

Matthew Arnold heard sadness in the roar of the waves on the shingle of Dover Beach, their 'tremulous cadence slow, and bring /

The eternal note of sadness in'.[10] A. E. Housman contemplated an untouched sandy beach, realising that to make any mark on it would be 'charms devised in vain' as the sea washes over all, 'effacing clean and fast'.[11] Tennyson hoped 'to see his pilot face to face'[12] when he sailed across a metaphysical sandbar. Melancholy shingle and sand but who cares for mud? Who looks at it and sees the sadness of what its loss really means?

Not all change is a loss. While saltmarsh forms across the temperate world, in tropical and sub-tropical regions it becomes replaced by mangrove forests – a sort of swamp version of saltmarsh. In a mangrove forest, trees instead of grasses grow out of the intertidal zone, their roots catching sediment and buffering the waves. Like saltmarsh, mangroves grow along the coastline and up rivers, wherever saltwater reaches. Climate change is encouraging them. In Florida, as far north as mangrove gets in North America, they have increased by 3,000 hectares.[13] This has been at the expense of saltmarsh, the warming conditions suiting the trees over the grasses. But it has led to an unclear ecological change: the benefits of the saltmarsh are also provided by mangrove. It has been suggested that they are better able to handle sea-level rise and more efficient at storing carbon than saltmarsh. We wait for studies to show the full effect.

We leave the marsh. Leave its wet and dry, its tangle of unwieldy words, its life forms familiar and alien, its pleasing and troubling associations. Walking back, the longest grass drags on our wellingtons, as if unwilling to let us go so easily. I thought I knew saltmarsh from years of my life spent looking out at it. Now I have realised that I didn't. Now that I have experienced it under boot and fingernail, I understand that the full saltmarsh experience is a vast thing, spanning water, tides, time; a myriad of moving parts that make up this most contrary of habitats. The only thing I know a bit better really is *this* saltmarsh, one among many, all different, all unique, all worthy of appreciation.

As we walk beside the marsh a wheatear scampers through the grass where we had been. The bird is running on long legs: pausing, running, pausing. With each pause the wheatear looks about, its eager eye examining the grass for insects. It turns away, its grey back tinged with a muddy brown that would have been missing earlier in the summer. The wing feathers, formerly black, are now fringed thickly with ginger. These are fresh feathers for the forthcoming migration: the flight to Africa, south of the Sahara. It is the first week of August and it feels like the height of summer is still to come but the evenings have been leaking light, the mornings slightly more chilled than before. The wheatears know this, know that August carries a tangible tang of autumn about it in southwest Scotland.

It is right to find a wheatear here. Marshes are places of transition and meeting. The birds of the uplands mingle with those of the lowland. Wetland encounters dry land, the submerged becoming the surface. And now summer is melding into autumn. There is the sense that everything here is about to intermingle and flow like water.

A sparrowhawk slips over the seawall and in a split second the sky is scattered teal and clattering wings: a three-figure flock of panicked ducks, snipe and starlings. From the seawall I follow the hawk's progress, fleeing fruitlessly from the scene, disappearing out of sight in the far corner.

August sunshine was half the story; I have returned in December to see the other half. My plan was to arrive an hour before the 10-metre high tide, but it has beaten me here, whipped by southerly winds, a seven on the Beaufort scale, gusting to eight; the path at the base of the sea wall is washed with unforgiving water the colour of concrete, the boardwalk that rises to the hide, where I am headed, partially submerged.

Necessity has beaten an unofficial path into the top of the sea wall, now a causeway between sea and freshmarsh. Water lies all around, birds are flying in all directions, fleeing for the higher ground, the dry ground, while there is still an hour to go until the water peaks. This morning the world is grey: low cloud, strong winds, squalling rain and the green of the marsh grass disappearing under the water, the forecast promising heavy rain later.

Saltmarsh comes alive on the high tide, the moon-pulled water charging this place with energy. Rain washes the hide window. The sparrowhawk returns: a juvenile hawk, brown, a female I guess from her large size, pursuing the starlings now. She flies far beyond, banks to face the flock and builds speed, shattering them into twisting ribbons of birds, coming back twice only for the flock to stay tight and alert to the danger. The third time she carries on going, buzzing past the hide, a thrilling blur of brown. The marsh is on high alert for the return of danger, everything hunkered down, vigilant.

After the fear, the lull of caution. A pause in the flurrying, frantic birds for the first time since my arrival. The hide window is glazed by salt and raindrops; the Cairnsmore of Fleet is hidden in low cloud, the lower flanks glazed in grey, and lorries flicker through trees as they rumble past on the A75. An elderly woman enters the hide. In the bay the wind churns the water into white-capped waves that dissipate on the unseen edges of the saltmarsh, silt and grass; *Salicornia* and *Spartina* diffusing the energy of the wind and water. A single great white egret flies across the bay, a rare winter visitor from the south, its glistening white plumage too pure, its long neck too delicate for a day such as this.

It doesn't take long for the caution to pass. The tide is a more pressing issue. The bay has changed and its charged birds have to deal with it. Lapwings and curlews beat the boundaries, looking elsewhere for dry land. Several hundred wigeon cluster in the calmer water of the marsh. A red-breasted merganser lurks in the channel of the Bladnoch. Shelduck, shoveler, mallard and teal are scattered across the water, lurking in the lee of the last patch of land that has yet to be covered. A flock of fifty redshank breaks off from there: five greenshank detach, circle back round and land on the breakwater, silver-feathered against the dull grey-brown of the water. A rock pipit, small and dark at this distance, scurries around their feet.

I point the greenshank out to the woman.

'Isn't nature wonderful,' she says.

'Aye.'

'Those thin legs keeping them up on a day such as this.'

An hour passes this way as if by accident: the birdlife buzzing, the water creeping until the fence posts marking the edge of the marsh disappear under the rising water. Today is particularly exceptional. I have been here before on 9-metre-high winter tides and witnessed the water taking over the marsh and nothing much happens; the birds capable, unflustered by the flood. That extra metre makes life difficult, another of the saltmarsh's stresses.

The land has been erased before my eyes. The green expanse of grass is gone, all that remains is a single wooden gate, its bottom two bars submerged, the seaward side of the fence sliding off under the surface. All is water now: an interregnum of the empire of mud. And it feels as though the hide has become cut off, a temporary, man-made island, our faith placed in the old wood and the unlikeliness of a freak wave. The marsh and sky reach out to each other in a blur of dirty brown-grey, the hills discernible only as smudges, the sea as shuffling white lines of distant breaking waves. And the moon, unseen, is having its effect on Earth, its gravity filling the bay with water like a cold, dirty bath.

The woman turns to me again.

'The tide here has been high enough to cover the top bar of the gate,' she says. 'One time a farmer over at Baldoon's shed door blew open and the sheep escaped. They swam up the river, got caught on the fence and drowned.'

And I don't know how to respond to that, other than to remind her that this tide has been used as a murder weapon before.

It happened in an arcane argument over God, in the seventeenth century, a period now known as 'the killing times'. Margaret McLachlan, sixty-three years old, and Margaret Wilson, eighteen, were tied to stakes driven into the mud of the old harbour, the area that is all saltmarsh now. McLachlan was driven into deeper mud as an example to Wilson, so she could see her fate and recant her covenanter faith. It did not work.

The spot on which they were both martyred is marked by a simple stone monument, lichened and weathered by life on the marsh. I have been there as the tide rolled up the creeks. It didn't feel right to stay there as the water rose above the banks.

A place such as this is impossible to know in its totality. Even to try we have to suffer the vertigo of deep time, delving into the sediment and the rock, searching through the layers of the past and the language of geology, all of which resists easy reading. We have to consider cordgrass and samphire and the constant flow of silt; the making and remaking of a landscape that can be reduced to a graph yet remains surprising and ever-changing. And while I shiver in the hide, counting curlews, considering the egrets, I realise that my understanding of this place is shaped by my own bias, my own opinion, by how I read the grammar of the weather and the words of the landscape. And I am haunted by the fact that someone else would read this all completely differently. Wigtown Bay has a slippery language, a spirit that wiggles through the mud like a *Corophium*.

And I love it. Fortunately so. For the tide remains high, the woman and I are still stuck in the hide, our island.

And so we watch the birds navigating between land, water and sky while we wait: the gulls hanging in the gale, the waders beginning to move again after the peak of the tide, their calls overlaying the dull roar of the wind. Soon they will be able to feed again, the bills turned low to the mud, long legs balancing their big bodies over thin feet, moving more elegantly than we ever can through the tricky mud; either delicately picking invertebrates by sight from the shining wet surface, or by feel; probing through the layers of mud with the motion of a sewing machine's needle, their bills soft and sensitive and able to feel for their buried food.

But we are not quite there yet. This is a delicately poised moment. We are all waiting for the moon to move, to reveal the mud again; for the life of a saltmarsh is a delicate interplay between land and water; sun and moon; bird and invertebrate; and us, marooned in the hide, waiting for it all to begin again.

Pre-dawn, brittle light, the wind whispering with curlews. The river still flows gently to the distantly lapping sea, but the channel is more mud than water, as if exhausted by the exertions of the tide. The pink in the western sky grows brighter. The only other movement in the moment, that of an egret picking its way over the mud. The only sound the distant mumbling of an engine. And then: apricot and amber. The rays break through, between banks of cloud. The far bank of the Bladnoch and its line of hawthorns are instantly thrown into black silhouette. There is nothing else I want to do but bask.

And on this Thursday at the dog-eared end of the year, it happens that I have nothing else to do, nowhere else to be, other than here. It's 8.45 a.m. and the sun has switched the day on and it begins to move around me. It starts with forty curlews beating their way upriver, a tight flock trailing a handful of stragglers, tag-alongs and slower birds; their undersides shining pale as they fly parallel to the light, turning black as they cut across it, following the curved-beak bend of the Bladnoch. More curlews drift into the marsh, gradually descending in flight, wings bowed, then straightening, curving down gradually, kicking out their legs for a gentle landing. They join the starlings on the grass that is no longer grey but glowing emerald in the morning sun.

Then silence. Overhead a peregrine falcon flies high into the light, fast and straight, its pale underside catching the sun's fire, this morning's meteor burning up through the sky. The passing presence of the peregrine tenses the saltmarsh up. The curlews are silenced. The starlings stationary on the grass. They take a beat. Adrenalin dips. Then the air fills with the sound of them, the bubbling calls again, the haunted whistles.

Here at the waterlands' end it is almost like a second cycle. Curlew calls ripple through air, as they did in the breeding season by the Clyde source, or in the Flow Country bogs, as they do in winter over the saltmarsh of the Solway. As the water lifts, drops, eddies, ceaselessly moving seawards, flowing like time, so does the wildlife, especially these curlews that move with the water, their calls that ripple and a flight that flows.

9

The Turning of the Tide

Water is not a story because stories finish. Water is like time, there is no end, not really, and no beginning, only a constant flow, a continuous cycle.

Heat, light, wind; the sea's water molecules are stirring, again. Energised, changing state, levitating like a magic trick, building into the sky. Ushered back towards land by the prevailing winds, the clouds hit hills or condense in cool air and the droplets released.

Raining again.

It falls on a changing world.

Conservation has two meanings.

The first is to preserve the old, the outmoded; keeping alive the things imperilled by time and progress, to keep things as they were.

This is great. But it is not the only way.

The other meaning is to improve, to reinvigorate; to try to lead the land and its habitats to a place where we can rekindle lost life; to encourage the new, strange and wonderful. Because why stay the same when the status quo led us into this mess?

This is the exciting, glamorous end of conservation. If politics is the art of the possible then conservation is the art of the impossible; each project a Lazarus act.

One such project is at Threave Garden and Nature Reserve,

run by the National Trust for Scotland. The reserve is an 81-hectare former dairy farm by the banks of the River Dee. You can't see the water from where I'm standing but you know it's there, snaking sinuously behind the built-up riverbank, or pooling before it, in the field's hollows. The grass here is thick and growing long in this late May sunshine. Other fields run with rivers of buttercups, but this one is washed all over with them, a golden slick swaying in the breeze.

As a farm, Threave was always a nice place to visit. A pair of fences funnelled you down from the car park, past cows and silage fields to the boggy ground by the riverbank and a view of Archibald the Grim's fourteenth-century castle on an island in the river. You would see ospreys in the summer, geese in the winter and roe deer all year around. But with a clear eye you would also get a sense of the farm as tired, the life of the place stretched thinly between its neatly fenced fields. When the tenant farmer relinquished the lease, instead of finding a new taker for the fields, the National Trust for Scotland had a different idea.

What has been done to a place can seem permanent but it doesn't have to be. The trust's new plan was to make the reserve a fully restored wooded wetland, self-sustaining and resilient to the globally warmed wetter future. Its work at Threave is an experiment in gradually undoing 300 years of change. To carry out its experiment it has a 100-year strategy: planning on an ecological scale. At the time of my visit, the trust is currently 3 per cent of its way through.

In 1986 the environmental historian Oliver Rackham wrote that 'the history of wetland is very largely the history of its destruction'.[1] That is still true today, globally. But the picture is slowly changing. In Britain, organisations such as National Trust for Scotland are taking action. Transfusions of life are happening across the country.

*

'The first change we noticed, I would say, was the longer grass,' says David Thompson, the head ranger for Threave nature reserve, his Yorkshire accent mellowed by decades in Galloway. 'Swifts and swallows are attracted to it, they're loving all this insect life in the field. This was a former silage field. Back in 2018 we ceased slurry and fertiliser spreading. That had a big impact. From then we started to see things change and the field revert back to . . .' he pauses for a second, juggling words in his head until the right phrase settles, '. . . to what it *wants* to be.'

You could call this a rewilding project. It is underpinned by natural processes and a respect for the idea that a field might actually want to be a certain way. I prefer to think of it as a rewetting project: the natural processes that are happening here are all underpinned by the most self-willed, if not completely wild, substance on Earth. Some 35 metres of man-made riverbank has been removed to let water in and 735 metres of drains have been filled to stop water heading out. Eight kilometres of fences have also gone. It is a grand encouragement for wildlife and water to go and flow where they want to, to intermingle, to let in and to hold on to what was excluded by the older style of management.

However you see it, rewilding or rewetting, the trust is keen to stress that its plan is a holistic one. It is guided by an appreciation of the series of connections that underpins nature, which is why we have begun talking about water by leaning against a gate, admiring the thick green and gold of the field, thinking about how taking the pressure off the land lets it rebound in extraordinary ways. Or, specifically, how a newly wet field changes the soil, awakens the dormant seed bank in the earth, which bursts forth with pollen, fuelling insects, attracting the swifts of the sky: the great cycles of life running to the rhythm of water.

The path takes us downstream first, through a boggy area by a filled-in drain; the first water that began staying put instead of flowing out. A faint sheen of nitrogen run-off pollution colours the surface of the water in the bright sunshine. It's not perfect yet. But black tadpoles wiggle to the surface then dive down. A handful of dace shoal like silver fingers, then vanish in the flick of a fin at a passing shadow. *Glyceria maxima*, reed sweetgrass, a flat green grass that looks like the stems of an iris, grows densely here in the wetter sections, a profusion of marshy life for the boardwalk ahead to guide you across.

The River Dee runs along the heart of Galloway, curving its way east from Loch Dee, flowing through the reservoirs at Clatteringshaws and Loch Ken, turning south then meandering past Threave to a hydro-electric power station at Tongland before slipping its way out to sea at Kirkcudbright. Dee is a common name for a river: there are five spread across Britain and Ireland (and four more with the name Don, which shares its etymology). It means 'river of the goddess'. Choose your goddess: they might be linked with fertility, beauty or wisdom.*

It is a wronged river. Its banks were built up in the seventeenth and eighteenth centuries to usher the water downstream and prevent the land that could be used for agriculture from flooding.† With the land no longer being farmed, the river wall became obsolete. In 2021, 25 metres of the bank upstream and 10 metres downstream were removed.

Through a gap in the hedgerow, the Dee can be seen where the bank has been removed. This is the exit, where the excess water in the wetland flows. We don't take the direct route to the entrance, instead meandering river-like around the control field. 'Nowt's going

* Other goddess rivers in Britain and Ireland: the Shannon, Boyne, Aeron and Severn.

† This was unpopular. In 1724 the Galloway Levellers smashed them down across the region. The landowners inevitably won as they did in the Fens.

to be done to this for five years; it won't be cut, nothing,' says David. Yellow rattle flowers stand proud of the grass, nettle-like leaves with folded yellow flowers held above. They parasitise the roots of grass, suppressing its growth, creating space for other plants to flourish. A scorpion fly sits in a tall buttercup, as if laird, surveying the downstream valley of green.

'OOH! Bloody hell!' David exclaims suddenly. 'Willow. Fantastic! Several.'

His colleague, Mary Smith, goes one better: 'And there's an oak there, look!'

'Oh wow!'

'And another oak!'

Their voices rise in increments with each gleeful surprise.

'Really loads of willow, that's amazing!'

David and Mary bounce off each other's enthusiasm. Their excitement is a tinderbox to my dry heart. The oaks are three apple-green leaves, poking out of a slender stem within the grass. The willow is more established. A foot of trunk, lime-leaved, standing up like one of the exclamation marks David and Mary speak in. Where small blocks of commercial conifer forestry stood on the site, they have been removed and replaced with native trees, tolerant of dampness and enabling other life to grow with them. Even where nothing has been done, the trees have found a way, via the very natural process: perhaps a jay that did not return to the acorns it cached under the soil in autumn, or spring's smoke of willow seeds, each one a spark catching.

The type of tree makes a difference. Deciduous trees are not as effective as conifers at transferring airborne pollutants to the water. Here, between fields and water courses, they intercept nitrate and other agricultural pollutants that run off the fields towards the watercourse. As their roots feel their way down through the soil, they create gaps and channels, allowing the soil to be more permeable than the agricultural land that was hardened and compacted by years of grazing, or by the shallow roots of conifer trees. Deciduous trees go some way to mitigating flooding, slowing the water's progress on its way to the river.

And when autumn comes, their leaves are whipped by the wind into the river, a natural release of nutrients. Their decaying energy gets used by the invertebrates that fuel the next generation of fish and waterbirds. Decaying branches and boughs of trees that fall into the water disturb the flow, causing patches of calm water, an important refuge for young fish, too weak to fight the faster current.

The trust has planted 2,000 alder, birch and aspen across the reserve. But each of these self-sown oak and willow saplings in the control field is proof of its faith in nature. By allowing these fields to be what they want to be, they'll be richer in life and better for the downstream water.

David told me that one of his wishes for Threave is for it to be a 'gateway' to nature. That gateway is increasingly crucial. Not just for the nature but for ourselves. Conservation organisations have been moving into fields they are not traditionally associated with, finding other ways to engage with people who may not have had much interaction with nature, who haven't felt the benefits of it. Since 2019, for example, the WWT has partnered with the Mental Health Foundation, researching 'blue prescribing' – the idea that time spent outside in wetlands and around water is beneficial for people experiencing poor mental health. In a project at Steart Marshes in Somerset, the WWT prescribed group walking, talking, poetry and weaving willow sculptures on their reserve. Their analysis after the project centred around 'social return on investment', a form of social accounting where intangible benefits are given a market value. It is vague – the report suggests we know the value of things such as a 'training course' – but their lead conclusion is stronger: every £1 that was put into the project led to £9.30 worth of benefit to the participants.[2]

It holds even if the participants are unable to walk or, like me, can't weave willow and dislike group activities. Using the method of ranking how photographs make you feel, a study from the University of Plymouth found an interesting nuance to the binary nature-good, urban-bad of other studies. The pictures they used were of blue spaces (water), green spaces (woods and parks) and built-up spaces, but on a spectrum – either only one of them or a combination of two. The first finding was that water in a built-up setting had a similar effect to a fully green space. The second finding was that the most valued pictures were of green and blue spaces together, leading the authors to conclude: 'The optimal environment may be the interface between land and large bodies of water.'[3] It sounds like a marsh to me.

Though this has perhaps not always been the case for all people. An American study from 1985 surveyed people's preferences for waterscapes, offering the choice of: mountain waterscapes; swampy areas; lakes, ponds and rivers; and large bodies of water. Mountain waterscapes were the most popular and swamps the least. The author states of swamps: 'It is apparently desirable to be able to figure out just where you are and how to leave quickly.'[4] The reasoning for this seems to be aesthetic: it was desirable then to be able to see a big view and be able to move easily through the waterscape. A study from 2021 drew links between familiarity and landscape preferences in North America; again it ranked swamps and marshes as both the least familiar and the least positively thought of environment types.[5]

Yet this stands in stark contrast to the work of the WWT's blue prescriptions. Getting people out into wetlands, breathing the air, seeing the water and the greenery, makes them happier, as does just looking at the wetland edge. This is the flipside, the unintended benefit to the work that the National Trust for Scotland is carrying out at Threave (where the paths and boardwalks have been made wheelchair- and pram-friendly – as accessible as it's possible for a path in a wild area to be). Access to these areas is important now, as it was 5,000 years ago when our oldest paths, the Post Track and Sweet Track of Somerset's Avalon Marshes, were made from oak planks pinned into place, it is thought for spiritual reasons as well as more practical ones.

David and Mary's desire for their landscape to be the gateway to nature is the first step. The work they've done and are doing seems to me to be the right way to go: the better the gateway, the more familiar people become, the more they like the environment, the happier they will be. The boom in life from the landscape becoming what it wants to be can only foster that connection, happiness growing like *Glyceria*.

Threave is thrilling. But there is another way to carry out this work, turning back 400 years of absence, if only we let it.

It is the cold bright light of an April morning. Eight of us strangers stand by a waist-height fence. We have dipped our wellies in disinfectant. The silver songs of spring spill down from the Sitka spruces lining the track – willow warbler, chiffchaff, blue tit – the frilly suns of coltsfoot, spangling the edges, where the rubble breaks up into boggy grasses. They are warmer than the sun that has no effect on a bitter lazy wind.

Cain gets the gate. Heather asks us, 'Have you ever been to a beaver wetland before?' One person says yes, the rest of us mumble no.

Heather Devey and Cain Scrimgeour run Wild Intrigue, a community-interest company running natural-history experiences and outreach in unusual ways. Our group of strangers is the first to experience this tour. Cain is softly spoken, hanging around at the back of the group, chipping in when he thinks it necessary. Heather is a polished speaker, full of natural-history knowledge from book and field. Together they had worked with beavers across the UK and Europe for a decade before the animals were returned to their home county of Northumberland.

We are in an obscure corner of the National Trust's Wallington Estate near Morpeth, a few kilometres from the manor house and manicured gardens that people usually visit. As at Threave, the Trust is trying to make its estate wilder, more biodiverse, and to that end it has fenced off 24 hectares alongside a fast-flowing burn. In July 2023 it introduced a family of beavers – two adults, two kits – relocated from the River Tay,* the first beavers in Northumberland since the fourteenth century.

Heather leads us down the track to a bridge over a tributary. The water is shallow, the channel narrow, though the water runs fast and clear. It carves its way through deep vegetation. It is a slow spring: the briars are sporadically coming into luminescent leaf. Birch, alder and willow in grey and brown grow by the bank, mapping the path of the water where the vegetation is too dense to see through.

'Who can be the first to spot signs of beaver?'

A few seconds later, we all point to a distinct felled tree at roughly the same time – an alder, the trunk gnawed to a sharp pencil-like point two feet up. Still connected to the stump, it angles down across the stream beyond 90 degrees, the remaining thin strip of soft wood unable to bear up the tree. Below it, beaten into the grass, a beaver's path.

This is where some mental gymnastics need to happen, because after half a millennium without beavers in the country, it looks bad. The idea of taking out trees is jarring, shorthand for environmental

---

* One of the conditions for beavers being allowed to live wild on the Tay is that if any landowner objects to beaver damage on their land, the animals can be relocated.

destruction. But deciduous trees can survive this. Heather explains coppicing to the group. We've known since the Stone Age that certain species of tree – oak, alder, hazel, willow included – if cut down at the trunk will regrow, with multiple thin straight trunks, useful wood for weaving, fences, furniture. It also allows light back into the understory, where flowers flourish in the new light and disturbed soil. It is a traditional but dying method of managing a woodland. A beaver on the riverbank is essentially carrying out this coppicing work for us. And if the tree does die, it doesn't go to waste: dead wood, a rare thing in our overly tidied landscape, is where fungi flourish; where the invertebrates who run the food chains feed and breed; refuges for fish if it lands in the river.

Beavers are guarded closely. They don't move far from the banks of the watercourse, so by the tributary the waist-height fence grows to shoulder height and tilts in at the top to prevent climbing; a layer of mesh is buried under the ground to prevent their tunnelling out. In the burn, metal bars like a prison let the water, fish and invertebrates through but keep the beavers inside their enclosure. It doesn't look brilliant – it is far from the wild ideal – but that's the politics of the beaver in England, the allowed realm of the possible, where they have to be supervised to stop them flooding an unsuspecting farmer's field or felling the wrong trees, behaviour that might not be appreciated. In Scotland they are allowed, under a looser sort of supervision, to exist in the wild. In England they are still subject to political hand-wringing, temporary licences that require renewal, allowed only in these enclosures; part prison, part laboratory. While these beavers are present in Wallington there are multiple studies under way to learn in greater detail their impact on this specific habitat, a flashy burn – meaning its level fluctuates wildly in response to rainfall – roughly 200 metres above sea level, surrounded by the rough ground of hill farms. There are wildflower surveys, breeding bird surveys, hydrology, riverflies, and even one to see if there is any impact on the white-clawed crayfish, our native, imperilled species of crayfish that can still be found in the catchment. Multiple conservation organisations, each with its own specialism, are all pitching in.

The walk carries on, skirting the conifers that the beavers will ignore. Marsh marigolds are beginning to bloom in the boggy ground, while meadowsweet leaves ramble through the grass – Cain tells me they love to eat those. We meet up with the main burn that runs through the site, its water russet with peat staining. In its flow a willow has been half felled, a tangle of roots and branches, beginning to green with the first leaves of its new life in the river. Cain points out that because the burn is flashy even the branches high up in the willow have the distinctive markings of a beaver's gnawing from when the water level was higher. Strands of dead grass have collected around the lower boughs of the trees, debris that the tree intercepts, slowing the water's downstream flow, lessening the likelihood of flooding. Below these boughs, gravel has collected too, forming miniature pools, where once this would have all been washed away.

Willow contains salicylic acid, one of the base components from which aspirin is made. This does not bother a beaver – in fact, as they can digest Japanese knotweed, bracken and fine-leaved water dropwort,* not a lot can poison a beaver. They are 'the ultimate vegetarian', in the words of beaver ecologists Frank Rosell and Roisin Campbell-Palmer, who go on to list 450 species that beavers have been recorded eating.[6] But they prize willow where it can be found. One of the quirks of willow is its propensity to reproduce vegetatively. That is, when a willow twig breaks off and floats downstream, if it gets stuck in the riverbank, it can easily put down roots and grow, a clone of the original tree. Most trees respond to a beaver

* Not as lethal as the other water dropworts but still substantially toxic.

feeding on them by growing further, rather than dying: the bitten edges have tight green curls of new leaves waiting to unfurl. It helps that the species they prefer to feed from, and unintentionally propagate and invigorate, are ecologically useful deciduous.

Further downstream Heather and Cain lead us along a bank above the burn. Under a stand of gorse on the far bank a lodge is being excavated: the beavers may be asleep in their burrows so we are hanging back from the water's edge, unwilling to cause any disturbance. It is where a guide helps because from here it looks like a messy pile of sticks behind a bush. They do admit it looks feeble.

We reach the riverbank a safe distance below. Looking back upstream, a dam is being constructed. Beavers want water for safety and food storage and so they attempt to hold on to it. It is impossible to hold on to water with just sticks, no matter how well you weave with willow; it is like cupping your hands together, water will slip through the smallest gaps that you can't close. But it has an impact regardless.

Due to the velocity of the stream over an exceptionally wet winter, the eighth wettest since records began in 1837,[7] the existing dams that they constructed last summer and autumn washed away. Some posts driven into the burn have given them a helping hand, catching flood debris and allowing the beavers to weave willow branches into them. It is incomplete. The far side doesn't yet reach the bank, allowing the water to ripple down and around, as well as through the small gaps in the structure. But it is the beginning of an effect: the water level is clearly higher up beyond it than it is below. With it even half-complete, the beavers are beginning to hold back the water. This enclosure is a landscape loaded with promise and potential for a better future.

When complete, there will be a series of dams down this burn, and a series of pools, where the water slows. Suddenly in this landscape where water has always rushed through, it will pause.

Computer modelling in North America predicts that in an undammed watercourse, water takes between three and four hours to travel 2.5 kilometres; with beaver dams it took eleven days for water to travel the same distance.[8] It works during storms, it works during droughts: whatever the extreme, a watercourse with beaver dams holds water better. They are the great ameliorating influence on the landscape, moderating the impact of a changing climate where the extremes will be sharpened.

In the recent past we copied their homework. In 2007 Pickering in Yorkshire flooded, causing £7 million worth of damage to the town. It was the most serious of a series of floods borne from intense rain in the moors above, overwhelming the becks and streams, the water cascading down into the town. The Forestry Commission built 167 leaky dams and a series of timber bunds – a barrier made from tree trunks – across the catchment upstream of the town. This was before beavers had become a mainstream tool in British conservation, when flood defences still consisted of walls and hard engineering. Yet their intervention of leaky dams and bunds – to all intents a human-made beaver dam, delaying the progress of stormwater – has reduced the annual chance of flooding in the town from 25 to 4 per cent.[9]

At Spain's Hall estate in Essex, eleven beavers have been living in a 4-hectare enclosure along a wooded stream for five years. In that time they have created nine dams, turning what was essentially a ditch into a swampy, broad wetland. Water now covers a quarter of the woodland, holding back what the Environment Agency estimates to be 3 million litres of water.[10] The neighbouring village of Finchingfield is on a different scale to Pickering: only eighteen properties are at risk of flooding so traditional methods of saving those houses from the heartbreak and devastation of flooding are uneconomical. The beaver has been an inspired solution. The villagers say they can feel the difference.[11]

*

I pick up a stick as we head back to the solid metal gates, running my finger over the smooth tooth-cut ridges. Iron is present in their incisors but still, I couldn't pick up a chisel and drive it through a tree and end up with something as soft and splinter free. It looks, more than anything, like the lines a knife makes through soft butter.

As I get the train home down the Tyne Valley, the track shadowing the river through woods, I feel a little bit rewilded too. My brain rationally knows that there are no beavers there. Yet each snapped branch, slumped trunk, eddying backwater looks full of possibility. There should be beavers there too. It just feels right.

Beavers are biological superchargers. Their presence on a river benefits otters, birds, some fish, insects and trees. They create diversity; their modifications to the landscape make it dynamic. They make it live, in the way that water has at Threave. What struck me most of all about my time in the enclosure is that beavers are the mammalian equivalent of water: wild, willed, redrawing the landscape around them. If land is bone and water is blood then beavers can – should – be the heart: keeping the water where it is needed; keeping the body alive.

They aren't right for everywhere. At Threave the river is too deep, the project too big: beavers aren't yet capable of removing 35 metres of riverbank in an excavator. Our historic shaping of the landscape takes more undoing. But, where possible, we should introduce beavers: rewilding the nation's small streams, encouraging them to hold on to water, instead of constantly ushering it downstream. It would have a major impact on flood reduction and biodiversity in the growing extremes of our flood-drought future.

Wallington was the second time I had seen signs of a beaver.

My first sign of a beaver came on the Tay, a surprise gnawed stump and wood chips next to the Birnam Oak. My third came a month after I'd been in the enclosure, walking down the Tay again, at dusk just north of Perth city centre. Light rain falling. A mayfly hatch persisting, the flies lekking into the darkening sky, the river rippling with fish and the air above flurrying with gulls. Light was running out. A Mallard quacked, a gull squawked at the same time, taking off upriver. My eye caught on a broader V of water, a skein of light. Through binoculars, drifting downriver where the gull and dark departed, the shape of a mammal in the river. It looked almost 2D. Flattened to the river surface, a gently raised head, snub-nosed; flat-backed and flat-tailed, and then gone. Disappeared into a glimmer of ripples against the black river. Even though it was here, semi-expected, it was a surreal thing. Put the binoculars down and the bank behind is Victorian mansions with manicured lawns leading down to wrought-iron benches overlooking the river. Something so wild in somewhere so suburban.

The next morning, light dulled by clouds. Further upriver, under a drifting osprey, along the foot-beaten path by the Tay's edge. Two steps between the bushes and suddenly, under a fallen tree at the river edge, that same face, same snub-nosed, startled-eyed staring from the water. A second later, it vanishes, a loud tail slap on the water a warning of my clumsy presence. A few metres downriver the water ripples over felled wood, mostly submerged. Reaching up out of the water, a spray of green willow leaves, living.

There is radical wetland conservation like Threave and then there's something more like stewardship, a careful shepherding of a place through all the damage that people might want to do to it.

On 26 July 2024, UNESCO declared that the Flow Country would become a World Heritage Site, a move that recognises the value of our waterlands, worthy of one of the highest accolades and protection; important beyond the borders of Britain for their scale and their species. It says that this blanket bog, our peat and water and acid and time, is as important to the world as the Great Barrier Reef or the waters of the Okavango Delta. It gives our bogs a global status and standing. In the unlikely event that a new threat appears – the Baird brothers, forestry or industrial peat-cutting reimagined for the twenty-first century – it will be heard internationally. As the first purely peat World Heritage Site, it should encourage others to care for their peat too. Having just won the status what happens next is an open question: to be answered not in months but years to come. Peatlands work on their own unique timescale after all.

It is not a perfect tool. World Heritage Sites can be delisted and many others are at risk. But in a world where shouting loudly counts, a potential global chorus of dismay and disgust at any threats is more powerful than not. It is a status that people innately understand better than a Ramsar listing. And in that respect it has worked almost immediately. A month after UNESCO's decision, one arms company asked for a licence to drop bombs by drones over the Strathmore peatlands, east of Forsinard, before revoking their request after they were told it was inside the World Heritage Site.

'Water is life' is a fine abstract statement but projects such as these give it flesh and bones, they make it real in a way we can immediately see. They have their tangible benefits but their power extends well beyond. They jolt us out of the potent rot of familiarity, make us look freshly at water again.

These projects also offer a picture of the future. That the land wants and needs to be wetter and we can welcome it in without devastation and flooding, if only we re-evaluate what we've done in the past in the light of an even wetter future. This is a vision of the waterlands – an oasis of life, carbon-storing, happiness-boosting – that holds across the dry spring of 2023 and the stormy wet winter of 2024. Allowing water onto land has many more benefits than draining and funnelling and trying to control water by brute force. And these are just two options. Reedbeds work as water purifiers as well as water and carbon storage. There used to be a running joke that the RSPB would turn everything into a reedbed if it could – now it doesn't seem so funny, just sensible.

Rivers need to run free and clean and the news reports an almost constant flow of bleak river headlines. During the 2024 Olympics and Paralympics, France spent €1.4 billion trying to clean the Seine sufficiently to host the swimming leg of the triathlon in the centre of the city, only for *E. coli* levels in the water to spike from sewage being washed out of their sewers by heavy rain. No athletes from either games could practise swimming in the Seine and the Olympic event was postponed for a day because of their antiquated sewer system – the same system, the same problem, the same heavy rain that we have in the UK.

I detect improvements though. When Labour won the 2024 General Election in the UK, a weighty section of its manifesto dealt with tackling the attention-grabbing issue of water companies and sewage pollution, although not with other forms of pollution and issues affecting British rivers and chalk streams. But what gives me hope for rivers is that this was an election issue at all. Because of campaigners like Fergal Sharkey, Windrush Against Sewage Pollution and Surfers Against Sewage raising awareness of the levels of water pollution in the country and the self-inflicted state of our water companies, which often cause it. Swimmers in our lakes and rivers are often those who shout the loudest, having put their bodies on the line and experienced the effects first hand.

It's not always possible to be there first hand. If it was, I would have been at November 2024's March for Clean Water. Instead, I was on childcare duty. So I had to watch the livestream of speeches the night after it happened, sorry not to have been part of it in person, feeling that communal energy of a shared outrage, marching on

Parliament Square, taking environmentalism to the heart of London with placards, costumes, slogans and speeches. I would have thought on the irony that the Thames threads its own concerns every second through the heart of London anyway, past the Houses of Parliament, easily ignored by the people who can't hear it speak or choose not to listen. Instead, 130 organisations and 15,000 people had to speak for it; organised by the charity River Action, whose chairman, Charles Watson, began with a moving litany of the names of the nation's rivers. Then a series of speakers: lawyers, artists, union leaders, celebrities, children; all powerfully talking along the same lines, telling everyone what they already know. That the state of our nation's waterbodies is a scandal. Water companies, intensive agriculture, the regulators were the rhetorical targets. The feelings were raw enough to coax thousands of people into the city on a gloomy winter's day; shared enough to make me dream that the brighter, cleaner future isn't far around corner.

Where there's a problem with rivers there's often the will to fix it. Compared with where they were in the nineteenth century, they are in an unimaginably good state – the salmon back in the Clyde, trout still abundant on the Test. But there is still plenty left that they need. Once we've sorted their needs we can think again about their wilder wants.

The Wildlife Trusts recently published a report on their adaptations to climate change. It featured a survey of their staff's threat perceptions alongside the issues they had in 2023 and the start of 2024 on their network of reserves. On first reading it seems contradictory: 90 per cent of Wildlife Trust conservationists consider drought to be a present threat to their reserves,[12] but the report goes on to detail the havoc caused by the wet winter, in which 130 per cent more rain fell than on average, noting that their peatland re-wetting project in Shropshire was delayed due to high rainfall leading to too much water.[13] But this is not a laughable contradiction – it is a symptom of the breaking climate; 2022 was the fifth driest year on record for the UK, part of a pan-European drought that was the continent's worst for 500 years; 2023 was the seventh-wettest year.

The world is going to become both drier and wetter. Fears of drought and floods at the same time are valid. As the climate warms it charges Earth's atmosphere with more energy. More warmth means more evaporation. The water cycle spins harder. More rain falling harder on a land that is increasingly more likely to be parched or puddled. Intense rainfall after a drought increases the chance of flooding as the land, hardened by time and temperature, is less porous. This is going to happen whether we make space for water or not; whether we let beavers weave willow to hold on to our water or not. We have always been adapting the water cycle but now it is not to our ends but to our detriment. The Mesopotamian and Indus Valley civilisations fell apart due to prolonged droughts; by 2050 it is thought that 4 billion people will be living in water scarcity.[14] In such conditions food becomes more expensive: wheat doesn't grow, dairy cows starve, people eat less and then their bodies are less able to deal with common diseases. These are conditions that turn people to violence. A three-year long drought in Syria, exacerbated by climate change, displaced around 1.5 million desperate people, which became a contributing factor to the country's brutal, bloody civil war.

It is ecology and geography that underpins all of this, that explains where water on Earth is and what it does and what we do to it. It is hard to imagine Britain without water – a world where the taps don't flow and the toilets don't flush – but unless things change, parts of southeast England are heading there. The jaws of death will clamp shut. Part of the reason why we are hurtling towards irrevocable climate change is this difficulty with picturing our lives affected this way; the everyday comforts we take for granted stripped away. Try to imagine your local landscape drained of its water: arid fields and riverbeds as dry as wadis; the only greenery in the foliage of the deepest-rooted, hardiest trees.

The Icelandic writer Andri Snær Magnason has seen this process happen in real time. Looking at home movies his grandparents made of their expeditions into their country's glacial heart, he admonishes his grandpa for filming the ice-stuck landscape instead of his grandma in her youth, rationalising the landscape won't change like she will age. Later he realises he is wrong. 'The glacier,' he writes, 'turned out to be as ephemeral as a person.'[15] A suitable scale for looking at things that are capable of dying. He goes on to predict that 'in the future, glaciers will be an alien phenomenon'.[16] Which I think is right. In the green south of Scotland, glaciers were the invisible hand that carved and swept and gouged out our landscape: it is impossible to imagine them back in their old places; nestling in the nooks between the peaks of the Lowther hills, as surreal as a sabre-toothed tiger walking down Dumfries high street. It might seem an extreme point, this, but it is an extreme that we should become familiar with. For in a world that is getting both wetter and drier, the forecast sharpens into extremes. Normality will be alien.

So we should be careful. Cautious in how we use water and all that entails – and not just in the length of our showers but what goes down our drains too – and keep pushing water companies and politicians for protections and improvements to our waterlands.

A willow tit flits down the hedgerow, appearing not to fly but to flick from willow to birch to hawthorn to willow; like a brown leaf driven by the breeze. Behind the willows, white sky merges seamlessly with the water of the Dee. Under the hedgerow a water rail shrieks as if complaining at its water-edge habitat shrinking, becoming all water and no edge.

December at Threave and it is as if time has turned the wetland tidal. After snow last weekend, the days that followed in the forecast held 100 per cent chance of rain. Torrential here, torrential there in the hills by Loch Dee, torrential in between where the water finds its way to the river. The rain is easing now on my head as I stand on the boardwalk by the Great Scrape.

It doesn't feel so long ago that this was all green and flowers and *Glyceria* and warbler song. I had returned once during summer, a fortnight after my first visit. Two weeks of oppressive heat, stretching into the upper twenties ('It's nae right,' passing strangers would say), the atmosphere so humid it felt as if you could swim through it. The only rain that had fallen was a sharp ten-minute shower of thick droplets that vanished almost as soon as they landed, slurped up by the parched earth. The early-cut silage fields had removed the green from the landscape, replacing it with a Mediterranean, dusty, pale brown. Southwest Scotland isn't supposed to be like that in June.

But Threave – glorious, wet, living Threave – had remained green, grasshoppers scratching sound into the dry air, the effervescent sedge warblers rising above and falling into shrubby willows, trailing their excited chattering song. The ospreys on their nest wobbled through the heat haze, the adults' bodies as parasols for their young.

Visiting Threave is like suddenly having fresh air to breathe. I brought other people here to look at what the landscape could be like; wet and green and alive in the heart of a hot summer, to see if water can gain a hold over them too. David Thompson asked me if I was inspired by being here and yes, Threave inspires me for the way that water has inspired the return of life to this old farm, like it could to so many tired farms across the world. New life to an old place, invigorating the old goddess.

Now, in winter, it is all grey water, cutting across the boardwalk that disappears below reflected clouds and the long strands of pink-footed geese that are passing overhead. The water is placid, ominous in the way that it submerges all but the handrails of the bridges and the highest dead sedges. It is possible to learn so many facts about water. To nail it down with science and say this is what it is. But facts never quite convey the feeling of it.

The torrential rain begins again, rings wrinkling the water. I need higher ground.

From the lower hide – which stands proud of the woodland and water, on perilous stilts – the water on Blackpark Marsh looks like the sea. Ducks are spread out as distant specks: mallard, wigeon and pintail but the vast flocks beyond dissolve into black dots. The geese are too far to identify with certainty. Whooper swan calls echo distantly like a wind-blown moan. Closer by, blackbirds are marooned on hawthorn islands. Others stalk where the water laps behind the hide in the woods, like waders, seeking out drowned worms. Somewhere in all this water, the Dee flows through its channel, hiding in plain sight among all this water. The rain drums its erratic beat on the hide roof. It gets heavier. It has just turned 1 p.m. but the light is already seeping out of the day minute by minute in an ecstasy of grey.

Higher up still, from the crest of the hill behind here, the complete 360-degree view of water can be seen. From east to west, where the river runs through the trust's land, it spools out, thickening, describing the essential geography of the fields, their dips and slopes and furrows, ignoring the hedges and few remaining fence posts that catch the eye anyway. One field of *Juncus* has become a perfect marsh where I don't recall one being. The only straight line in the landscape is the field that borders the neighbouring farm's raised track. To the west, where the Dee curves an elaborate S, the horizon is all water, shining grey, the brightest thing left in the day until the farm over the valley turns its arc lights on. Earlier, rows of hills were visible beyond that farm; now in the shroud of rain it is as far as I can see. The fields are only getting wetter.

At the car, I peel my coat from my skin. Dump it with my hat and bag, everything sodden through. It is a thrilling thing, Threave. Where the washland of the Fens is an admission that the land can't be fully drained, its straight edges a sign of human control, the dead hand of a human plan, here it feels organic. A land shaped and changing by the rhythms of water; going with its flow. This would have been catastrophic for the reserve a few years ago in its previous guise. Now it is brilliant for it. A practical demonstration that working with water negates the problems of working against it.

As the water cycle keeps turning, rain is water's beginning and its end. We should not shrug it off and curse it but pay attention to it. As water and land are indivisible, so is rain: we might as well fall back in love with it, to consider what it wants and needs. Because between water's beginning and its end, between the blood and bone of it, we find the heartbeat of the planet.

Epilogue

There have always been waterlands in my life. Before my toddler stumble-trip into the cold chalky water of the River Cam, there was being carried around Wicken Fen, feeding ducks in the pond, being beside Lake Windermere, glimpsing bogs and reeds from the car window. But maybe, in the real beginning, there was saltmarsh.

In the photo we think I am two years old. This is all we have to go on. The single negative that my dad scanned and emailed to me this morning contains no other information.

I am standing next to my brother. We are wearing wellies (a wise move) and blue corduroy trousers (perhaps less so). A too-big hand-me-down coat hides most of me. Zoom in and you'll see my buck teeth through my open mouth, my lower jaw hanging slack with befuddlement. I'm holding a pair of plastic binoculars and looking the wrong way, my eyes peering through the lenses, the eyepieces angled down at the photographer's knees, which must have seemed confusingly small and far away.

Behind me that pale blue sky of the past stretches out to the horizon. It fades out into a dark green line, where the definition is lost in the blur of the film. It could be pine, grass or farmed field; it looks low and flat, one of those wide-open East Anglian horizons I grew up with. Winter sedges, brushed by the wind, lay bleached at my feet. And all the square kilometres between the sedges beneath and the horizon lie below us, the brindled green and brown saltmarsh vegetation; salt-slicked grasses and the sublime of open space. The sort of day that still dazzles three decades later.

That space: it could be anywhere along the coast between Lincolnshire and Essex. But I'm not sure that being any more specific about the *where* matters so much as the *what*.

I wonder if Dad knew the effect that being there would have on me. For my early childhood was one walked out along straight raised paths, by flat fields of space, almost certainly in the teeth of an East Anglian breeze, a constant background exposure to fen and saltmarsh. I am too young in the picture to know what any of this means but I have the desire to copy Dad. To whip my head around as he does as our noisy family party startles meadow pipits, swept from the sedges like leaves in the breeze; to leave an ear for the myriad whistles of waders that I will begin to learn properly as a teenager; to keep an eye trained on the horizon, waiting for a bird of prey to break the cover of the solid and soar in that brilliant sky, circling over the land that comes and goes with the water; that grows long grass and strange lives. Another cycle, compelled to be repeated, the gifts of being by saltmarsh given to another generation.

If water is blood and land is bone then I feel the waterlands deep inside me, part of me, the water in my brain and heart and muscle flowing within the chalk of my skeleton. Water still holds me. I am still drawn to be near it. And as my journey with water has completed one spin around the cycle, I feel it more than ever: the desire to watch the rain, walk along the river or to a loch, to be still beside a bog or marsh, open to the animating spirit of life and all it holds. It is as strong now as it must have been when I was two, looking the wrong way through a pair of binoculars, a small being beside a vast world of water.

Acknowledgements

The first words of this book I wrote four years ago. As it takes a village to raise a child, it takes a village to raise a book from an abstract idea to this thing that you hold in your hands; from the hands of editors, to people who find the time to answer questions, reply to emails or say encouraging things online. Those are the little boosts that help so much when writing happens in lunch breaks or into the night or on my phone in the stray moments snatched from the routine of a day.

Time, when writing, functions much the same as it does while raising a child: the past four years have simultaneously flown and inched past; flowed smoothly and juddered by in chaos, internal and external. Through it all, A, I've been there for you and you've been there for me; you've been teaching me while I've been trying to teach you the names of things, and reminding me of what's important when the sentences just won't come (and when they won't stop). And so, with all my heart and mind and love, this book is for you (I don't expect you to read it).

Thanks must go to my agent, James Macdonald Lockhart, who took my inchoate ideas and helped them make sense, helped coax it to a proposal when I was sleep-deprived and helped usher it through to completion with his buoyant emails.

At Elliott & Thompson all my thanks and appreciation go to Pippa Crane, editor par excellence, Amy Greaves in publicity and

Instagram Sam. And to Jill Burrows the copyeditor, Marie Doherty the typesetter and Emma Rogers the jacket designer.

My thanks to Adrian Turpin of Wigtown Book Festival for commissioning the Saltmarsh Library Project, from which Chapter 8 has been culled, a project that was funded by NatureScot's Year of Coasts and Waters. Little snippets of that commission have popped up elsewhere across the book as well, because, well, water always flows. I also wish to thank the Author's Foundation at the Society of Authors for their grants for works in progress and also to the Royal Literary Fund for presenting me with the JB Priestley Award as well. In a bleak outlook for writers actually getting paid to make books, the work and generosity of the SOA and RLF are vital and this past year wouldn't have worked without them.

The past few years also would not have worked without Miranda Cichy – parenting co-pilot – my thanks for the freedom you let me have for when I needed to be alone in a field thinking of water, for the dicey driving down unsuitable roads and for swimming on my behalf and then telling me about it.

The list of people to thank who helped me when disaster struck is too long to include here. But my thanks and gratitude in particular have to go to Jill Asher and the Asher family for providing shelter to a broken writer. Thanks to Victoria Cichy for reading drafts of a few early chapters, also David Borthwick, Rhian Davies, Milly Revill Hayward, the Hewitt family, cousin Lizzie, Adam Murphy, Nick Patel, Mary Smith, Rebecca Tanner, Elizabeth Tindal, Dave Thompson and Charles 'Trees' for their assistance with fieldwork, expertise with places and species, and time generously given to me. Researching this book reminded me of Barry Lopez's line, 'The astonishing level of my ignorance confronted me everywhere I went.'[1] I felt similarly. Any mistakes are my own.

Not everything that gets researched makes it into the finished, final thing. So for the research I did under the project that didn't make it into the book, my thanks and apologies to Rowena Chambers of the RSPB, Liam Templeton of ARC, Ruedi Nager of Glasgow University, Rona McGill and Jason Newton of the SUERC.

And because no one else would read this far down an acknowledgments page, thanks to my parents for believing in me.

Endnotes

Introduction: Before the Fall

1. Gavin Maxwell, *A Reed Shaken by the Wind* (London: Eland, 2003), p. 55.
2. Alok Jha, *The Water Book* (London: Headline, 2016), p. 95.
3. bell hooks, *Belonging* (New York: Routledge, 2009), p. 1.
4. Graham Swift, *Waterlands* (London: Picador, 1992), p. 9.
5. Figures from https://earthobservatory.nasa.gov/features/water.
6. Ismail Serageldin, 'Water Wars? A conversation with Ismail Serageldin', *World Policy Journal*, vol. 26, no. 4 (winter 2009/2010), pp. 25–31.
7. https://www.theguardian.com/environment/2022/sep/12/us-west-megadrought-climate-disaster.
8. https://edition.cnn.com/2023/04/12/weather/florida-flash-flood-fort-lauderdale/index.html.
9. https://www.theguardian.com/australia-news/2022/dec/30/australias-ninth-wettest-and-20th-hottest-year-looking-back-at-the-weather-in-2022.
10. https://www.globalwaterforum.org/2022/07/27/how-many-australians-lack-safe-and-good-quality-drinking-water.
11. https://www.gov.uk/government/news/lack-of-water-presents-existential-threat-says-environment-agency-chief.
12. https://www.unesco.org/reports/wwdr/2023/en.
13. https://www.usgs.gov/media/images/endpoint-colorado-river-mexico
14. T. C. Smout, *Nature Contested* (Edinburgh: Edinburgh University Press, 2000), p. 90.
15. https://www.wwt.org.uk/our-work/threats-to-wetlands.
16. Joan Didion, 'Holy Water', in *The White Album* (London: 4th Estate, 2017), p. 59.
17. Nan Shepherd, *The Living Mountain* (Edinburgh: Canongate, 2011), p. 27.
18. Tristan Gooley, *How to Read Water* (London: Sceptre, 2017), p. 1.

1 Unreliable Sources

1. Robert Chambers, *Popular Rhymes of Scotland* (Edinburgh: W. & R. Chambers, 1858), p. 26.
2. https://www.bbc.co.uk/news/uk-scotland-39139549.
3. Both quotations from Nan Shepherd, *The Living Mountain*, p. 4.
4. Sir Walter Scott, *The Betrothed* (London: John C. Nimmo, 1894), p. xvii.
5. Definition from the Dictionaries of the Scots Language: www.dsl.ac.uk.

2 Actions and Reactions

1. Jamie Linton and Jessica Budds, 'The hydrosocial cycle: defining and mobilizing a relational-dialectical approach to water', *Geoforum*, vol. 57 (November 2014), pp. 170–80.
2. Ibid., (p. 177).
3. Charles Perfect et al., *The Scottish Rivers Handbook* (Aberdeen: CREW, 2013), p. 22.
4. https://www.latimes.com/environment/story/2023-01-27/colorado-river-in-crisis-agriculture-under-pressure.
5. https://www.latimes.com/california/story/2024-04-28/san-joaquin-valley-official-accused-of-epic-california-water-heist.
6. https://scholarlycommons.pacific.edu/cgi/viewcontent.cgi?article=2477&context=mlr, pp. 321–2.
7. Chambers, *Popular Rhymes of Scotland*, p. 26.
8. https://www.marxists.org/archive/marx/works/1880/soc-utop/ch01.htm.
9. https://hakaimagazine.com/features/the-hatchery-crutch-how-we-got-here/.
10. Barry Lopez, *The Syntax of the River* (San Antonio: Terra Firma, 2022), pp. 14–15.
11. Richard Muir and Nina Muir, *Rivers of Britain* (Exeter: Webb & Bower, 1986), p. 24.

3 The Pure, the Wholesome, the Insulted

1. Hugh MacDonald, *Rambles Round Glasgow* (Glasgow: John Cameron, 1860), pp. 83–4.
2. Marion Bernstein, 'A Song of Glasgow Town', in *Radical Renfrew: poetry from the French Revolution to the First World War*, edited by Tom Leonard (Edinburgh: Polygon, 1990), p. 304.
3. T. M. Devine, *The Scottish Nation: A Modern History* (London: Penguin, 2012), p. 168.

4. Ibid.
5. E. Ashworth Underwood, 'The History of Cholera in Great Britain', *Proceedings of the Royal Society of Medicine*, vol. 41 (November 1947), pp. 168–9.
6. John Burnet, *A History of the Water Supply to Glasgow* (Glasgow: Bell & Bain, 1869), p. 26.
7. Underwood, 'The History of Cholera in Great Britain', p. 170.
8. https://www.nationalarchives.gov.uk/education/resources/victorian-industrial-towns/cholera-in-glasgow/.
9. https://www.poetryfoundation.org/poems/50578/a-description-of-a-city-shower.
10. https://www.bbc.co.uk/news/uk-scotland-58040852.
11. 'The River Clyde', *Nature* (30 November 1876), pp. 99–100.
12. Stephen Mullen, *Glasgow Slavery Audit* (March 2022), p. 31. Published online: https://www.glasgow.gov.uk/media/7180/Glasgow-Slavery-and-Atlantic-Commerce-An-Audit-of-Historic-Connections-and-Modern-Legacies/pdf/Glasgow_Slavery_Audit.pdf?m=1711362464620
13. https://it.wisnae.us/.
14. Joseph Conrad, *Heart of Darkness and Other Tales* (Oxford: Oxford University Press, 2002), p. 105.
15. Charles More, *The Industrial Age: Economy and Society in Britain 1750–1985* (Harlow: Longman, 1989), p. 116.
16. Devine, *The Scottish Nation*, p. 571.
17. Ibid., p. 592.
18. Ruth H. Thurstan and Callum M. Roberts, 'Ecological Meltdown in the Firth of Clyde, Scotland: Two Centuries of Change in a Coastal Marine Ecosystem', *PLoS One*, vol. 5, no. 7 (July 2010), p. 12.
19. Ibid.
20. Kiri Rodgers et al., 'The legacy of industrial pollution in estuarine sediments: spatial and temporal variability implications for ecosystem stress', in *Environmental Geochemistry and Health*, vol. 42, no. 4 (April 2020), p. 1064.
21. C. H. Vane, et al., 'Polyaromatic Hydrocarbons (PAHs) and Polychlorinated Biphenyls (PCBs) in Surface Sediments from the Inner Clyde Estuary, U.K.'. Published online: https://nora.nerc.ac.uk/id/eprint/990/2/Baseline_PCB_Clyde_short_versionpdfa.pdf.

4 Still Life

1. https://dgnhas.org.uk/biography/sir-william-jardine-applegarth.
2. Ibid.

3. World Lake Database: https://wldb.ilec.or.jp/Lake/EUR-28.
4. 'Lake Tanganyika: Experience and Lessons Learned Brief'. Published online: https://www.ilec.or.jp/wp-content/uploads/pub/22_Lake_Tanganyika_27February2006.pdf.
5. Brian Moss, *Lakes, Loughs and Lochs* (London: William Collins, 2015), p. 33.
6. https://www.gov.uk/government/publications/state-of-the-water-environment-indicator-b3-supporting-evidence/state-of-the-water-environment-indicator-b3-supporting-evidence.
7. Lord Byron, 'Don Juan', in *Romanticism: An Anthology*, ed. by Duncan Wu (Malden: Blackwell, 2006), p. 933.
8. Primo Levi, *The Periodic Table*, trans. by Raymond Rosenthal (London: Abacus, 1986), p. 128.
9. Moss, *Lakes, Loughs and Lochs*, p. 180.
10. Ibid.
11. Ibid.
12. Ibid., p. 59.
13. https://www.bbc.co.uk/news/uk-england-cumbria-67557310.
14. https://www.bbc.co.uk/news/uk-northern-ireland-66835897.
15. Roger Deakin, *Waterlog* (London: Vintage, 2000), p. 3.
16. Davide Filingeri et al., 'Why wet feels wet? A neurophysiological model of human cutaneous wetness sensitivity', *Journal of Neurophysiology*, vol. 112, no. 6 (June 2014), p. 1460.
17. Robert Angus Smith, *Air and Rain* (London: Longman's, Green and Co, 1872), p. 225.
18. Ibid., p. 229.
19. Ibid., pp. 225 and 266.
20. Michelle L. Bell and Devra Lee Davies, 'Reassessment of the lethal London fog of 1952: Novel Indicators of Acute and Chronic Consequences of Acute Exposure to Air Pollution', *Environmental Health Perspectives*, vol. 109 suppl. 3 (July 2001), p. 393.
21. Derek Ratcliffe, *In Search of Nature* (Leeds: Peregrine Books, 2000), p. 111.
22. R. N. B. Campbell, 'Tail deformities in brown trout from acid and acidified lochs in Scotland', in *Acidification and Fish in Scottish Lochs*, edited by P. S. Maitland et al., (Grange over Sands: Institute of Terrestrial Ecology, 1987), p. 65.
23. Malachy Tallack, *Illuminated by Water* (London: Transworld, 2022), p. 28.
24. Ibid., p. 67.
25. On the planet b612 podcast: https://www.youtube.com/watch?v=0n34VPM9UtQ (25:30 minutes in).

26. Rick Battarbee, 'Forestry, "Acid Rain", and the Acidification of Lakes' in *Nature's Conscience: The Life and Legacy of Derek Ratcliffe*, edited by Des Thompson et al., (King's Lynn: Langford Press, 2015), p. 389.
27. https://www.gov.scot/publications/key-scottish-environment-statistics-2016-9781786525505/pages/2/.
28. Ed Rowe, et al., 'Air Pollution Trends Report 2024: Critical Load and Critical Level exceedances in the UK', p. 3. Published online: https://uk-air.defra.gov.uk/assets/documents/reports/cat05/2502271309_Air_Pollution_Trends_Report_2024_V2.pdf.
29. Michael Wigan, *The Salmon* (London: William Collins, 2015), p. 191.
30. Derek Ratcliffe, *Galloway and the Borders* (London: William Collins, 2007), p. 283.
31. The UK Forestry Standard: the government's approach to sustainable forestry, p. 119. Published online: https://cdn.forestresearch.gov.uk/2023/10/The-UK-Forestry-Standard.pdf.

5 Green as a Dream

1. Rupert Brooke, 'The Old Vicarage, Grantchester', in *The Faber Book of Landscape Poetry*, edited by Kenneth Baker (London: Faber, 2000), pp. 409–10, ll. 11–12.
2. Charles Rangeley-Wilson, 'Catchment Based Approach: Chalk Stream Restoration Strategy 2021, Main Report', p. 25. Published online: https://catchmentbasedapproach.org/wp-content/uploads/2021/10/CaBA-CSRG-Strategy-MAIN-REPORT-FINAL-12.10.21-Low-Res.pdf.
3. Kent Buse and Kate Bayliss, 'England's privatised water: profits over people and planet', *BMJ*, vol. 378 (23 August 2022), p. 1.
4. https://www.theguardian.com/environment/2024/feb/26/british-irish-rivers-desperate-state-pollution-report-trust#.
5. https://unearthed.greenpeace.org/2023/07/31/sewage-uk-water-pollution/.
6. Ciara Marie Fitzpatrick, 'The Hydrogeology of Bromate Contamination in the Hertfordshire Chalk: Double-Porosity Effects on Catchment-Scale Evolution', doctoral thesis, p. 267. Published online: https://discovery.ucl.ac.uk/id/eprint/1306703/1/1306703.pdf.
7. https://wildfish.org/blog/englands-water-supply-shortfall-set-to-be-4-8-billion-litres-per-day-by-2050-and-the-water-companies-main-solution-cutting-demand-looks-like-pure-fantasy/.
8. Alfred, Lord Tennyson, 'The Brook', in *The Faber Book of Landscape Poetry*, pp. 39–41, ll. 25–36.
9. https://www.theguardian.com/uk-news/2018/feb/19/country-diary-should-diarists-meddle-in-politics-1968.

10. Karen Bakker, 'Neoliberalizing Nature? Market Environmentalism in Water Supply in England and Wales', *Annals of the Association of American Geographers*, vol. 95, no. 3 (2 November 2010), p. 549.
11. https://www.theguardian.com/money/2023/dec/18/water-firms-use-up-to-28-percent-of-bill-payments-to-service-debt-in-areas-of-england#:~:text=As%20of%20March%2C%20English%20water,of%20company%20accounts%20for%202023.
12. David Hall, 'Water and Sewerage Company Finances 2021: Dividends and Investment – and Company Attempts to Hide Dividends', PSIRU Working Paper, p. 15.
13. Deakin, *Waterlog*, p. 32.
14. Ibid, p. 33.

6 Under the Skin

1. Mark Cocker, *Our Place* (London: Vintage, 2019), p. 240.
2. Robin Crawford, *Into the Peatlands* (Edinburgh: Birlinn, 2018), p. 15.
3. https://iucn.org/resources/issues-brief/peatlands-and-climate-change#3846.
4. https://www.nature.scot/doc/scotlands-national-peatland-plan-working-our-future.
5. Edward Maltby, *Waterlogged Wealth* (London: Earthscan, 1986), p. 9.
6. https://www.npws.ie/peatlands-and-turf-cutting.
7. https://www.youtube.com/watch?v=KjymNcM_8Ok.
8. https://news.google.com/newspapers?id=x7dAAAAAIBAJ&sjid=yqUMAAAAIBAJ&pg=3300%2C3975869.
9. J. A. Baker, *The Peregrine* (London: Collins, 2010), p. 23.
10. Peter Friend, *Scotland* (London: Collins, 2012), p. 409.
11. https://www.fcrt.org/the-flows/.
12. Ratcliffe, *In Search of Nature*, p. 205.
13. Derek Ratcliffe, *Bird Life of Mountain and Upland* (Cambridge: Cambridge University Press, 1990), p. 165.
14. Lean and Rosie, cited in *Nature's Conscience: The Life and Legacy of Derek Ratcliffe*, edited by Des Thompson et al., (Kings Lynn: Langford Press, 2015), p. 407.
15. R. J. Payne et al, 'The future of peatland forestry in Scotland: balancing economics, carbon and biodiversity', *Scottish Forestry*, vol. 72, no. 1 (2018), p. 35.
16. Both quotes cited in Thompson et al. (eds.), *Nature's Conscience*, pp. 445 and 444.

17. Ibid., p. 415.
18. Ibid., p. 430.
19. https://whc.unesco.org/en/about/.
20. Fairlie Kirkpatrick Baird et al., 'Projected increases in extreme drought frequency and duration by 2040 affect specialist habitats and species in Scotland', *Ecological Solutions and Evidence,* vol. 4, no. 3 (August 2023), pp. 1–12.

7 Rebirth

1. Francis Pryor, *The Fens* (London: Head of Zeus, 2019), p. 42.
2. Graham Swift, *Waterland* (London: Picador, 1992), p. 11.
3. https://www.literarynorfolk.co.uk/Norfolk%20Poems/powte's_complaint.htm.
4. Cocker, *Our Place*, p. 184.
5. Tom Williamson, *An Environmental History of Wildlife in England 1650–1950* (London; Bloomsbury Academic, 2013), pp. 102–3.
6. Edward Storey, *Spirit of the Fens* (London: Robert Hale, 1985), p. 75.
7. Thomas E. Dahl and Gregory J. Allord, 'Technical Aspects of Wetlands: History of Wetlands in the Coterminous United States', United States Geological Survey, Water Supply Paper 2425, 1999.
8. Aldo Leopold, *A Sand County Almanac: and Sketches Here and There* (Oxford: Oxford University Press, 1968), p. 162
9. M. W. Lang et al., 'Status and Trends of Wetlands in the Conterminous United States 2009 to 2019', U.S. Department of the Interior; Fish and Wildlife Service, Washington, D.C. (2024). Published online: https://www.fws.gov/sites/default/files/documents/2024-03/wetlands-status-and-trends-2009-2019-signed.pdf.
10. Ibid., p. 6.
11. Leopold, *A Sand County Almanac*, p. 162.
12. https://www.hrw.org/legacy/backgrounder/mena/marsharabs1.htm.
13. Andy Brown et al., 'Bitterns and Bittern Conservation in the UK', in *British Birds*, vol. 105, no. 1 (January 2012), pp. 58–65.
14. Norman Sills and Graham Hirons, 'From Carrots to Cranes', in *British Wildlife*, vol. 22, no. 6 (August 2011), p. 389.
15. Hans Brix et al., 'Are *Phragmites*-dominated wetlands a net source or net sink of greenhouse gases?', *Aquatic Botany*, vol. 69, nos 2–4 (April 2001), pp. 313–24.
16. Kai Whitaker et al., 'Vegetation persistence and carbon storage: Implications for environmental water management for *Phragmites australis*', *Water Resources Research*, vol. 51, no. 7 (June 2015), pp. 5284–300.

17. https://www.bto.org/understanding-birds/birdfacts/marsh-harrier.
18. Tarjei Vesaas, *The Hills Reply*, translated by Peter Owen and Elizabeth Rokkan (New York: Archipelago Books, 2019), p. 52.
19. Cocker, *Our Place*, p. 74.
20. https://www.lincstrust.org.uk/what-we-do/wildlife-conservation/projects/bourne-north-fen.

8 Between Land and Water

1. Richard Mabey, *Flora Britannica* (London: Sinclair-Stevenson, 1996), p. 98.
2. Clive Chatters, *Saltmarsh* (London: Bloomsbury, 2017), p. 115.
3. T. A. Haynes, 'Scottish Saltmarsh Survey National Report', *Scottish Natural Heritage Commissioned Report No. 786* (2016), p. 95.
4. S. Q. An et al., 'Spartina Invasion in China: Implications for Invasive Species Management and Future Research', *Weed Research*, vol. 47, no. 3 (May 2007), p. 185.
5. W. Austin et al., 'Blue Carbon Stock in Scottish Saltmarsh Soils', *Scottish Marine and Freshwater Science*, vol. 12, no. 13 (September 2021), p. 19.
6. Laura Airoldi and Michael W. Beck, 'Loss, Status and Trends for Coastal Marine Habitats of Europe', *Oceanography and Marine Biology: An Annual Review*, vol. 45 (June 2007), pp. 345–405.
7. P. Esselink et al., 'Wadden Sea Ecosystem No. 25: Quality Status Report 2009, Thematic Report No. 8', Common Wadden Sea Secretariat (CWSS), 2009, p. 3.
8. Xinxin Wang et al., 'Rebound in China's coastal wetlands following conservation and restoration', *Nature Sustainability*, vol. 4, no. 12 (December 2021), p. 1076.
9. Michael McCarthy, *The Moth Snowstorm* (New York: New York Review of Books, 2015), pp. 66 and 78.
10. Matthew Arnold, 'Dover Beach', in *Victorian Poetry*, edited by Francis O'Gorman (Malden: Blackwell, 2004), pp. 312–13, ll. 13–14.
11. A. E. Housman, *Collected Poems* (London: Penguin, 1995), p. 207, ll. 25 and 23.
12. https://www.poetryfoundation.org/poems/45321/crossing-the-bar.
13. https://floridadep.gov/rcp/rcp/content/floridas-mangroves.

9 The Turning of the Tide

1. Oliver Rackham, *The History of the Countryside* (London: J. M. Dent & Sons, 1989), p. 375.

2. https://www.wwt.org.uk/uploads/documents/2024-01-30/wwt-sroi-of-blue-prescribing-evaluation-final-report-updated.pdf.
3. M. White et al., 'Blue Space: The Importance of Water for Preference, Affect and Restorativeness Ratings of Natural and Built Scenes,' *Journal of Environmental Psychology*, vol. 30, no. 4 (December 2010), p. 490.
4. Thomas R. Herzog, 'A Cognitive Analysis of Preference for Waterscapes', *Journal of Environmental Psychology*, vol. 5, no. 3 (September 1985), p. 238.
5. Giancarlo Mangone et al., 'Deciphering Landscape Preferences: Investigating the Roles of Familiarity and Biome Types', *Landscape and Urban Planning*, vol. 214 (October 2021), pp. 1–14.
6. Frank Rosell and Roisin Campbell-Palmer, *Beavers: Ecology, Behaviour, Conservation and Management* (Oxford: Oxford University Press, 2022), p. 140.
7. https://www.theguardian.com/uk-news/2024/mar/01/february-was-warmest-on-record-in-england-and-wales-met-office-says.
8. Cited in Rosell and Campbell-Palmer, p. 307.
9. Tom Nisbet et al., 'Slowing the Flow at Pickering: Final Report', published by Defra, May 2015.
10. https://www.gov.uk/government/news/five-years-of-beaver-activity-reduces-impact-of-flooding.
11. https://www.spainshallestate.co.uk/nfm_beavers
12. Kathryn Brown et al., *Embracing Nature: Climate Change Adaptation at the Wildlife Trusts*, p. 9. Published online at https://www.wildlifetrusts.org/sites/default/files/2024-08/Embracing%20Nature_0.pdf.
13. Ibid., p. 12.
14. Chunyang He et al., 'Future global urban water scarcity and potential solutions', *Nature Communications*, vol. 12, no. 1 (August 2021), p. 2.
15. Andri Snær Magnason, *On Time and Water* (London: Serpent's Tail, 2020), p. 175.
16. Ibid., p. 183.

Acknowledgements

1. Barry Lopez, *About This Life* (London: The Harvill Press, 1999), p. 139.

Index